FUTURE FOODS

Edited by **Heimo Mikkola**

Future Foods
http://dx.doi.org/10.5772/65132
Edited by Heimo Mikkola

Contributors

Pierre Rougé, Dushimirimana Severin, Thierry Hance, David Damiens, Weiping Yin, Junna Liu, Zhenjun Sun, Hao Jiang, Afam I. O. Jideani, K. R. Netshiheni, Noe Aguilar Rivera, Régulo Carlos Llarena-Hernández, Christian Michel-Cuello Michel-Cuello, Martin Roberto Gámez- Pastrana, Teresita De Jesus Debernardi-Vazquez, Marcello Nicoletti, Paola Del Serrone, Maria João Fraqueza, Luis Patarata, Mai Kodama, Hidemitsu Furukawa, Masaru Kawakami, Jin Gong, Azusa Saito, Hideaki Tamate

Published by InTech
Janeza Trdine 9, 51000 Rijeka, Croatia

Publishing Process Manager Romina Rovan
Technical Editor SPi Global
Cover InTech Design team

First published September, 2017
Printed in Croatia
Legal deposit, Croatia: National and University Library in Zagreb

Additional hard copies can be obtained from orders@intechopen.com

Future Foods, Edited by Heimo Mikkola
p. cm.
Print ISBN 978-953-51-3551-7
Online ISBN 978-953-51-3552-4

For EU product safety concerns:
IN TECH d.o.o., Prolaz Marije Krucifikse Kozulić 3, 51000 Rijeka, Croatia,
info@intechopen.com or visit our website at intechopen.com.

Meet the editor

Heimo Mikkola has received his PhD degree in Applied Zoology and Limnology from the University of Kuopio, Finland. He has worked over 30 years abroad in Africa, South America, and Central and Southeast Asia mainly with the Food and Agriculture Organization of the United Nations (FAO). As the resident representative of FAO in Africa and South America, he has overseen a number of mushroom farming and beekeeping projects and done private studies on the use of wild foods. After retiring from UN /FAO, he has been a part-time visiting professor in the Kazakh National Agrarian University in Almaty lecturing lately to 32 university teachers on Green Biotechnology and Global Food Security, including the insect farming as future potential in the food and feed production of the world. He has written almost 600 work reports and scientific papers, including 10 own books, mainly on applied zoology and owls, from all continents of the world.

Contents

Preface

World population growth has led to global search for new protein alternatives. Now we are feeding well over seven billion people with 8.4 billion tons of food including 42 kg of meat and 19 kg of fish (as average per capita food availability). Net food availability is one ton/person/year as the food loss and waste take 1.3 billion tons/year out of the total. Farmed animals for food consumption include 20 billion chickens, 1.0 billion cattle and 0.8 billion pigs, 1.2 billion ducks, 1.1 billion sheep, 1 billion goat, 0.8 billion rabbits, and 70 million tons of food fish (aquaculture). From our globe, only 40% of ice-free land surface is suitable for agriculture, when forest takes 30%, deserts and mountains take 25%, and the urban and wetlands take the rest [1].

From the total agriculture land, 70% is used for feeding livestock and 30% of all grains produced are fed to animals (80% of soy and half of the corn production). Similarly, 70% of the total fresh water is used for agriculture with 110 million tons of chemical fertilizers and 2.3 million tons of pesticides (like one-third of glyphosates—roundup), and agriculture is causing 14–17% of total greenhouse gas emissions and producing 6 billion tons of manure [1].

When the world population doubled during the last 50 years, the agriculture production tripled, and the increase caused only 12% addition in farmed area. But the water consumption was twice as fast as population growth.

It is anticipated that by 2050, we will have nine billion people to feed—how can we manage? That time 60% of people will be living in urban areas, and they will have higher demands for animal-derived proteins up to 60 kg of meat and 30 kg of fish. What comes to farmed animals for human consumption, there is a need for 7% annual increase in chicken and aquaculture to produce 60 billion chickens and 140 million tons of aquaculture products.

Other protein alternatives will be needed and could according to [1] come from (a) capturing more out of oceans (medusa, jellyfish, krill); (b) farming the sea (macro- and microalgae (spirulina—high-tech); (c) artificial proteins (Ap) but presently in vitro meat costs 120,000 USD/kg; (d) more out of agro-industry processing (Aip) by-products (corn gluten, brewers' grains, yeasts, potato protein concentrate, distiller's dried grain with soluble); (e) farming more plant protein sources (oil seeds, legumes, forages, trees (like *Moringa* leaves)); (f) compete for land, water, fertilizers, farm inputs; or (g) high-capital/high-tech (Aip, Ap, spirulina algae) potential (regional niche markets).

We also should worry about how to feed our animals: in 2013, it was estimated that 795 million tons of cereals, which is one-third of cereal production, went for the feed [1]. And by 2050, we need an additional 520 million tons for our animals, i.e., half of the estimated cereal production. Monogastric sector (chickens, pigs, aquaculture) consumed in 2013 155 million tons of mainly soy feed protein. In 2050, an additional 110 million tons of feed protein are needed, this being half from cereals/soy protein sources. In 2013, some 110 million tons of course grains were used for bioethanol production.

As scarcities of agricultural land, water, forest, fishery and biodiversity resources, as well as nutrients and nonrenewable energy are foreseen, insect rearing as "mini-livestock" is a solution for food and feed security [1]. However, insect farming is not only a modern invention as it has been practiced past 5000 years by cultivating silk moth (*Bombyx mori*), bees, carmine scale insects, like Polish cochineal (*Porphyrophora polonica*), and lac insects like lac beetle (*Kerria* or *Laccifer lacca*). Insect rearing for pet food, fish bait, zoos, medicinal purposes, and the ornamentals started in the early 1900s. These insects included some 20 species from mealworms, fly larvae, butterflies, crickets, and beetles. Since the 1960s, some 100 insect species have been used for the integrated pest management and plant pollination. Insect farms for the human food started around 2000 by using less than ten species like crickets, mealworms, etc. By 2010, mass rearing of insects began as feed for farmed animals with less than five species like fly larvae—black soldier fly (*Hermetia illucens*).

What are the nutrition values? Insects have good but highly variable nutritional values depending on species, metamorphical stage, habitat, and diet. Processing of the insects plays also a significant role in determining nutritional content. Insects are particularly important as food and feed supplements compared with fish. They provide satisfactory amounts of energy and protein and meet amino acid requirements. Insects have plenty of monounsaturated and/or polyunsaturated fatty acids and are high in micronutrients like biotin, copper, folic acid (in some cases), iron, magnesium, manganese, pantothenic acid, phosphorus, riboflavin, selenium, and zinc.

Depending on the source of insect and type of animals, the insect meals can replace 25–100% of soymeal or fish meal in the feed of our domesticated animals or fish. Aroma, taste, and carcass quality will not be different with the use of insect meals. Overall levels of essential amino acids are good in insect meals, over 42% in black soldier fly larvae to 61% in silkworm meal and over 57 to 82% in defatted insect meals, and the protein digestibility is good. Insect meal energy value is higher than those of both soymeal and fish meal, and the oil content is as high as 36%. Insect meals are low in calcium, except black soldier fly larvae, which is 7.6% [1].

Insect farming contributes to a closer, local circular economy in livestock rearing as locally produced sidestream from agriculture, agro-industries, food, and waste management is available to local insect farmers to produce proteins and fats. These can be used as ingredients for livestock, meat, and fish producers in the same region. This will shorten the chain for feed producers by incorporating better locally produced ingredients. Insect rearing can improve local farming economies, including small farm operators.

Benefits of insect farming can be summarized as follows [1]: (a) it can provide food for two billion people, plus natural feed of chickens, carnivore fish in aquaculture, and pigs; (b) insects are a good source of protein, fats, minerals, plus micronutrients; (c) feed conversion efficiency is four times better than in cattle rearing; (d) it can be fed with organic waste, so it is not competing with foods for humans; (e) it has fast breeding cycle (10 days to 1 month); (f) it requires less water, land, greenhouse gas emissions; (g) it may help reduce the use of antibiotics in feed; (h) it is socially more inclusive (easy to farm, no need for huge investment, or "high-tech" skills); (i) it opens opportunities for business and jobs at any level, everywhere in the world; (j) it supports local-circular agro-economy; and (k) it is interdisciplinary (food-feed-health-waste disposal-climate change and jobs).

However, now one of the largest barriers to increasing consumption of edible insects, as well as their use as animal feed, is legislation. In my own country, Finland, production of insects is legal only for the own use but not for commercialization. It is also interesting that for highly valued free roaming chicken insects are part of their feed, but insects are not legally allowed in chicken feed in the European Union countries.

My own interest in insect food started when living some 22 years in Africa, east, south, and west starting 1977 in Kenya and ending 2004 in Nigeria.

In Kenya, we had an old cook from Kakamega where he surely ate insect larvae, termites, etc. However, he was dead worried when he saw the Louisiana red swamp crayfish (*Procambarus clarkii*) sample I had captured for a party. For my cook, the alien crayfish was a dangerous insect nobody he knows would eat [2]. For us "musungus" in Nairobi it was a delicacy and good serve for a Scandinavian type of crayfish party. Cook was not at all surprised to see people get sick the morning after the party but without understanding that this morning sickness was because of too many hard drinks rather than the crayfish eaten. This just shows how sensitive issues any new food items can be in the diverse cultures.

When living in Malawi (1993–1997), I got even more involved in the use of insects in the diet of local people. In Malawi, the population growth outstrips food production, especially during times of drought. According to my small survey [3], the use of wild foods is often the only coping mechanism during the malnutrition and famine. Wild fruits, mushrooms, nuts, seeds, and leaves were eaten by 83–97% of the interviewed families (110 in total). Insect food was almost as popular: larvae/caterpillars 68 %, grasshoppers/locusts 46%, flying ants/termites 86%, lake flies (*Chaoborus edulis*) 17%, crickets/cicadas 58%, and other insects 58%.

In this book, we have nine chapters ranging from mushroom farming to smart packaging and 3D printing of future foods.

Even the sky is not the limit for the future foods, which can benefit largely from future space travel of food design. Edible algae, bacteria, fungus, and even viruses have already been tried as food sources and produced in the laboratories for the space menu. That type of food can later be produced also in the spaceship despite of limited space and water availability. If programmed correctly, single-celled organisms could produce the basic elements of everything that people need in space, from screwdrivers to hamburgers [4]. This new technique combines bacteria with 3D printing. If you need a screwdriver, you use a Petri dish filled with common bacteria (*Bacillus subtilis*) that can make bioplastics. In a bioreactor, where the sugar-producing cyanobacterium (*Synechococcus elongatus*) is already present, the bacteria use sugar to produce bioplastics. Finally, this stuff can be put in a 3D printer to make a screwdriver. Ready-made hamburger is not as simple, but the bacteria can produce the different carbohydrates, fats, and vitamins. People in the spaceship will need to finalize the edible burger in the kitchen.

I want to thank the Publishing Process Manager Romina Rovan for her time-consuming efforts to get all the authors to deliver their chapters after the corrections. Without her active attitude, some important chapters would have been left out from the book.

Heimo Mikkola
University of Eastern Finland,
Finland

References

[1] Vantomme, P. 2015. World population growth and the search for new protein alternatives. Conference Paper https://www.researchgate.net/publication/282753885

[2] Mikkola, H. 1978. Ecological and social problems in the use of the crayfish, *Procambarus clarkii* in Kenya. Freshwater Crayfish 4: 197-206.

[3] Mikkola, H. 1997. The use of wild foods in Malawi. The Society of Malawi – Historical and Scientific. 50(2): 40-53.

[4] Larsen, J. 2016. Space bacteria will produce dinner for Mars astronauts. Science Nordic 3 p. http://sciencenordic.com/space-bacteria-will-produce-dinner-mars-astronauts. Accessed 21/06/2017.

Chapter 1

Competitive Edible Mushroom Production from Nonconventional Waste Biomass

Noé Aguilar-Rivera,
Régulo Carlos Llarena-Hernández,
Christian Michel-Cuello,
Martin Roberto Gámez-Pastrana and
Teresita de Jesús Debernardi-Vazquez

Additional information is available at the end of the chapter

http://dx.doi.org/10.5772/intechopen.69071

Abstract

Worldwide, there is an increase in acreage dedicated to the competitive production of fruits, vegetables, flowers, and palms crops as productive diversification for food, feed, fiber, and fuels. However, in developing countries, there is malnutrition by an inadequate diet caused by deficiency in quality or quantity of food. Therefore, options are needed for the production of foods, mainly of high protein content such as edible mushroom from by-products. In Veracruz, Mexico, there is a large megadiversity of wastes derived from endemic plants, fruits, legumes, pods, leaves, straws, and flowers that are generated in a large amount and are disposed off through open-field burning without a specific use. The objective was to evaluate the potential of 30 nonconventional by-products and wastes for the production of low-cost oyster mushrooms *Pleurotus ostreatus*. Biological efficiency (BE) varied from 17.65 to 180% and at least the 60% of the evaluated substrates (BE greater than 50%) are viable for the production of mushroom *Pleurotus* especially in view of its low contamination in trials and abundance and availability and diversity throughout the agricultural year as wastes. Besides, the spent substrates were converted into organic manure compost, vermicompost and bocashi to close the cycle for new food production.

Keywords: protein-rich food, lignocellulosic by-products, competitive rural enterprises, oyster mushroom, organic manures

1. Introduction

1.1. Issues and constraints for the competitive and sustainable food production in Mexico

According to FAO estimates [1], 795 million people worldwide lack the quantity and quality of food needed to enjoy an active and healthy life. The vast majority live in developing regions. To better understand the problem, the concept of food and nutritional security defined by the Food and Agriculture Organization of the United Nations (FAO) has been established as "Physical and economic access to sufficient safe and nutritious food to meet their needs food preferences in order to lead an active and healthy life." This concept began to take on importance since the 1920s, with efforts to mitigate hunger; In the 1950s and 1960s, government actions were directed toward increasing the productivity and efficiency of marketing systems [2]. At the global level, this concept became more important as a result of the financial, energy and food crisis of 2008. Major advances have been made since the early 1970s in increasing food or raw material production, but this progress slowed down during the 1990s, due to factors such as climate change and climate change routing of crops for the production of biofuels.

In accordance with Urquía-Fernandez [3], food security is based on four elements: availability, access, food use, and stability of supply; to these, we must add two effects of an inadequate intake of foods that are obesity and malnutrition. That is why although in Mexico the average energy availability is 3145 kilocalories per person per day, 14% of the infant's present malnutrition, 30% of the adult population suffers obesity in different degrees and more than 18% of the total population is classified as food poverty by income. One of the reasons for this phenomenon is explained by the processes of development and sociocultural changes that are occurring in Mexico, which directly influence the eating patterns characterized mainly by a high intake of foods rich in cholesterol, saturated fats, sugars, and salts [4]. In addition [5], indicates that products oriented toward a more economical market usually have the highest content of fats, sugars, and additives. Poor nutrition has social and economic effects, a hungry person is not an efficient worker, they present more frequently illnesses and a poor nutrition of mothers has repercussions on children with lower physical and mental capacity [6, 7].

The current global capacity to produce food is such that the needs of all its inhabitants can be met. Numbers from the National Institute of Statistics, Geography, and Informatics (INEGI) indicate that there is sufficient food availability in Mexico [8]. FAO established an indicator called Food Energy Supply (SEA), defined as the daily consumption of food energy per person over a given period and including food produced or imported by countries for human consumption; for Mexico in 2000, this indicator has a value of 3159 kcal/day and its energy needs per person are calculated at 2182 kcal/day; therefore, the information indicated is validated. Unfortunately, there are factors that have limited adequate access to food by the population. These factors include transfer costs and the lack of price regulation of the basic food basket [9].

Currently in Mexico there is a very noticeable change in the dietary habits of the population, since the patterns of consumption have left their local and regional characteristics to gradually evolve to a standard national, that is to say the traditional Mexican food has succumbed to the search of the imitation of the food model of the developing countries. An example of this is the change in food with abundant cereals and legumes, a variety of fruits and vegetables and small portions of food of animal origin, which has been displaced by a diet based on products of animal origin with their corresponding intake of saturated fats and cholesterol, very refined cereals with low fiber content and high consumption of sugars and sweeteners. The product of these changes in diet is reflected in chronic diseases such as malnutrition, overweight, obesity, and diabetes [10].

Food is a physiological need characterized by food intake influenced by various factors such as culture, food, and economic availability, and environmental and family issues. The human metabolism needs to cover a daily requirement of approximately 50 nutrients to maintain their health, some of these nutrients are considered essential because the organism can not produce them, so it is necessary to ingest them in the daily diet [11]. There are studies that indicate that 90% of cases of malnutrition are caused by an inadequate diet, which can be caused by a deficiency in quality or quantity of food eaten [12]. Currently, in Mexico, it is estimated that 1.5 million children under 5 suffer from chronic malnutrition. As for overweight and obesity, this is caused by the abuse of high-calorie foods, with high concentrations of fats, salt, and sugars and with low amounts of vitamins, minerals and fiber, coupled with little or no physical activity; Genetic factors also contribute. In Mexico, 10% of children under 5 years of age and 70% of the economically active population present cases of overweight and obesity [13]. Diabetes, a disease considered the epidemic of the twenty-first century, since in 2007 its prevalence in Mexico was 7% [14].

The change of the patterns of feeding in Mexico has been given since the conquest with the introduction of new species of cattle and other species of cereals; however, some rural communities and certain sectors of the urban population maintain part of that pre-Hispanic food culture that is characterized by foods with great nutritional value. Before the Spanish conquest, the feeding of the different pre-Hispanic Mesoamerican cultures was based on a great diversity of animal and vegetal species peculiar to each of the geographic regions (both wild and domesticated). The availability of products used as food at this time was very varied; among them, we can mention edible plants, herbs and condiments, flowers, and fruits and highlight the use that was given to mushrooms named generically as nanácatl or forest fungus (**Table 1**). Estrada [15] points out that up to six names are registered in the Florentine Codex to refer to edible fungi. Pre-Hispanic cultures appreciated their great biological diversity and developed forms of sustainable use of their resources; One example of this is the multiple benefits they obtained from the main food of the indigenous peoples, maize: the sweet stalk was consumed directly, the leaves were used to wrap food like *tamales* where it served as protection during cooking, *elote* which is consumed directly desgrained or whole, from these grains are obtained *tortillas*, sweet *tamales* and drinks; Even pests of corn are exploited as are the worm (*gusano elotero*) and the fungus that infestated them called *cuitlacochin*. With the voyages of discovery and commercial exchanges with the world, many species were introduced and successfully adapted due to their use by indigenous groups, and many of these associated species survive to this day and constitute almost half of the dietary in México [16]. In the period 2000–2013, it is estimated that the Mexican population reduced their consumption of fruits and vegetables by 30% and consumed 40% more sweetened drinks and 10% more carbohydrates in their diet [7].

Common name	Náhuatl name	Scientific name
Calabaza	Ayutli	*Cucurbita pepo*
Bean	Etl	*Phaseolus* spp.
Chile	Tzilli	*Capsicum annum* y *C. frutescens*
Corn	Cintli	*Zea mayz*
Chayote	Chayutli	*Sechium edule*
Huazontle	Quauzontetl	*Chenopodium berlandieri* ssp. *nuttalliae*
Quelite	Quilitl	*Chenopodium album*
Quintonil	Quintonilli	*Amaranthus hybridus*
Purslane	Itzmiquilitl	*Portulaca oleracea*
Red tomato	Xitomatl	*Lycopersicum esculentum*
Green tomato	Tomatl	*Physalis philadelphica*
Nopal	Nopalli	*Opuntia* spp.
Vanilla	Tlilxochitl	*Vanilla planifolia*
Achiote	Acyotl	*Bixa orellana*
Chipilín	Chipilli	*Crotalaria longirostrata*
Epazote	Epazotl	*Chenopodium ambrosioides*
Colorín	Zompantle	*Erythrina* spp.
May flower	Cacaloxóchitl	*Plumeria* spp.
Izote	Iczotl	*Yucca elephantipes*
Custard apple	Zacualtzapotl	*Annona cherimola*
Chicozapote	Xicotzapotl	*Manilkara zapota*
Guava	Xalococotl	*Psidium guajava*
Tejocote	Texocotl	*Crataegus mexicana*
Capulín	Capolin	*Prunus serotina*
Nanche	Nantzinxocotl	*Byrsonima crassifolia*

Table 1. Pre-hispanic food consumed in some regions of Mexico [15].

The agricultural systems in México have been oriented to intensive production characterized by various factors such as the implementation of technologies to increase productivity, an adaptation of crops to the demands of international markets and genetic modifications to improve product varieties. In addition, government support has been received with the aim of promoting the productive reconversion and diversification of traditional crops. Besides, irrigation is poor; therefore, Mexican agriculture is still mostly rainfed and highly affected by climatic variations (**Figure 1**).

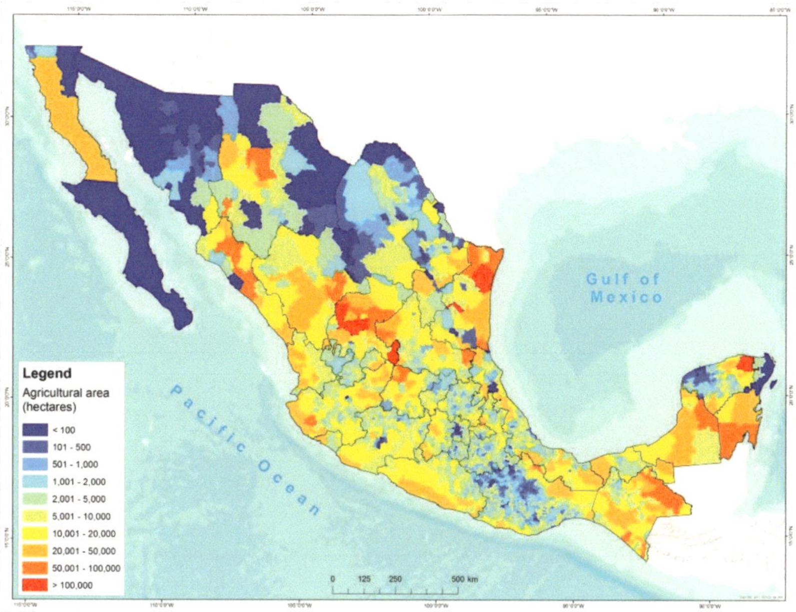

Figure 1. Rainfed agriculture in Mexico [8].

Agricultural activities produce significant amounts of waste (**Table 2**) and by-products that can be used in the formulation of animal feed, edible mushroom and in the production of compost and vermicompost, techniques that consist in the degradation of organic matter by the action of microorganisms and worms respectively, to produce humus or organic manure, products that increase the availability of plant elements [17]. Unfortunately, it is estimated that only 50% of the total produced is used. The burning of agricultural waste is a widespread practice in Mexico to dispose of or reduce the volume of agricultural products, and it is an economical and easy practice. According to conclusions published by CEC [18], it is estimated that biomass burning "Produces 40% of carbon dioxide (CO2), 32% of carbon monoxide (CO), 20% of particulate matter or (PM) and 50% of polycyclic aromatic hydrocarbons (PAHs) emitted to the environment on a global scale." A very important alternative for the use of agricultural residues is the use of biotechnology that allows the establishment of solid state fermentation of fibrous residues in the production of edible mushroom and later from their waste, animal feed and compost.

Taking as reference the report of the global condition of agriculture and food [20] and based on projections of population growth, world food demand in 2050 is expected to be at least 60% higher than 2006. Taking this stage into account, it is necessary to develop processes that are orientated toward sustainable food production. On the one hand, adequate food must be ensured for future generations, and on the other hand, these processes should no longer affect natural ecosystems and their ecological services; we must even go beyond restoring and creating the natural resources that have been damaged.

Crop	Generation coefficient
	(t residue/t product)
Wheat	1.20
Barley	1.35
Oats	1.35
Rye	1.35
Corn	2.00
Rice	1.50
Sorghum	1.70
Citrus fruits	0.15
Sunflower	2.00
Cotton	2.00
Sugar cane	1.50

Table 2. Biomass generation coefficient in different crops [19].

2. Potential for competitive food cultivation from biodiversity in Veracruz, Mexico

Veracruz State is a crescent-shaped strip of land located in Eastern Mexico wedged between the Sierra Madre Oriental to the west and the Gulf of Mexico to the east. Its total area is 78,815 km^2. It extends about 650 km north to south but its width varies from 212 to 36 km. Even when the elevation of a great portion of the State is between sea level to one thousand meters, the topography in the west in some areas changes drastically, creating a complex system of mountains and valleys which can reach altitudes over five thousand meters.

The combination of these factors generates an interesting mosaic of climatic conditions according to the methodology of Ref. [21]. In Veracruz, annual mean temperature goes from almost 27°C (81°F) to −1°C (30°F). The lowlands located in the south are relatively warmer than the lowlands in the north. In contrast, the central portion of the State has temperatures that can reach 0°C in the highest mountains. Annual precipitations show too a great variation. Even when about 40% of the territory has precipitations between 1000 and 2000 mm, in some western areas, yearly rainfall is below 500 mm. In the other extreme is possible found regions where the annual rainfall is over three thousand mm (reaching a maximum of 3230 mm). The distance to the coastline, the elevation range, and the physiographic conditions also generates enormous variations in moisture, temperature annual range, and temperature seasonality besides the variations in the distribution of rainfall. The environmental conditions present the Veracruz permit the existence of different biomes (**Figure 2**).

http://dx.doi.org/10.5772/intechopen.69071

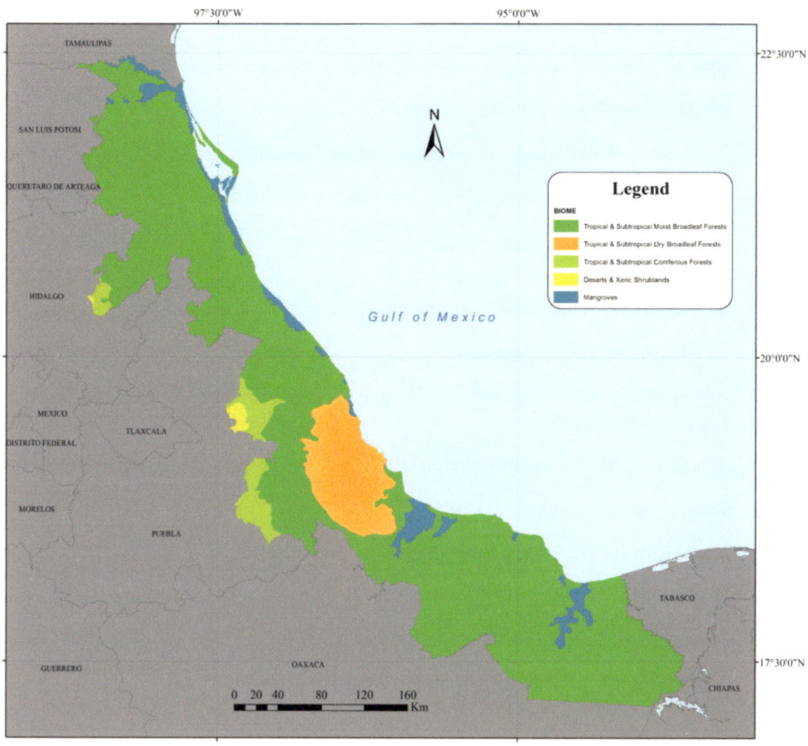

Figure 2. Biomes in Veracruz.

Although tropical & subtropical moist broadleaf forest occupies a great portion of the State, tropical and subtropical dry broadleaf forest, tropical and subtropical coniferous forest, xeric scrubland, and mangroves are present too. Consequently, the biodiversity existing in the territory is one of the greatest in the country. Current data indicates that native vascular plants species are over 5472 species, i.e. 28% of the total number of species present in Mexico and 2% of the vascular plants that exists in the world.

The climatic mosaic present the Veracruz combined with the edaphic variability also had allowed the cultivation of a great variety of native and exotic crops (from annual herbaceous to tropical and subtropical tree species), which found the appropriated conditions to grow adequately. As a result of the distribution of mountain ranges in Veracruz and their latitudinal positions each region on the state has contrasting soil types, climatic, hydrologic, and biotic characteristics (**Table 3** and **Figures 3–5**).

However, despite the high potential for the production of various crops, Veracruz is the third state in Mexico with the highest level of poverty, therefore, alternative food production required to complement traditional systems of food production and transit to sustainability.

Major soil group	Characteristics	%
Acrisols	Soils with subsurface accumulation of low activity clays and low base saturation	8.5
Cambisols	Weakly to moderately developed soils	7.0
Rendzinas	Dark soils rich in organic matter over calcareous material	4.4
Gleysols	Soils with permanent or temporary wetness near the surface	5.1
Phaeozems	Soils with a thick, dark topsoil rich in organic matter and evidence of removal of carbonates	12.5
Lithosols	Azonal soil consisting chiefly of unweathered or partly weathered rock fragments	2.5
Fluvisols	Young soils in alluvial deposits	0.4
Kastanozems	Soils with a thick, dark brown topsoil, rich in organic matter and a calcareous or gypsum-rich subsoil	0.2
Luvisols	Soils with subsurface accumulation of high activity clays and high base saturation	9.6
Nitosols	Soils with shiny surfaces on structural faces (peds) of the soil	0.5
Arenosols	Sandy soils featuring very weak or no soil development	1.4
Regosols	Soils with very limited soil development	14.8
Andosols	Young soils formed from volcanic deposits	5.8
Vertisols	Dark-coloured cracking and swelling clays	25.5
Planosols	Soils with a bleached, temporarily water-saturated topsoil on a slowly permeable subsoil	0.2
Solonchaks	Strongly saline soils	0.2
Others		1.4

Table 3. Soil types from Veracruz [22].

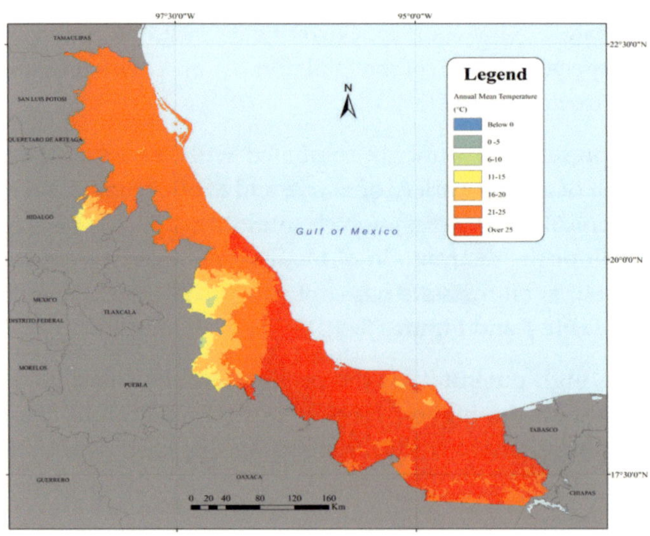

Figure 3. Average temperatures in Veracruz.

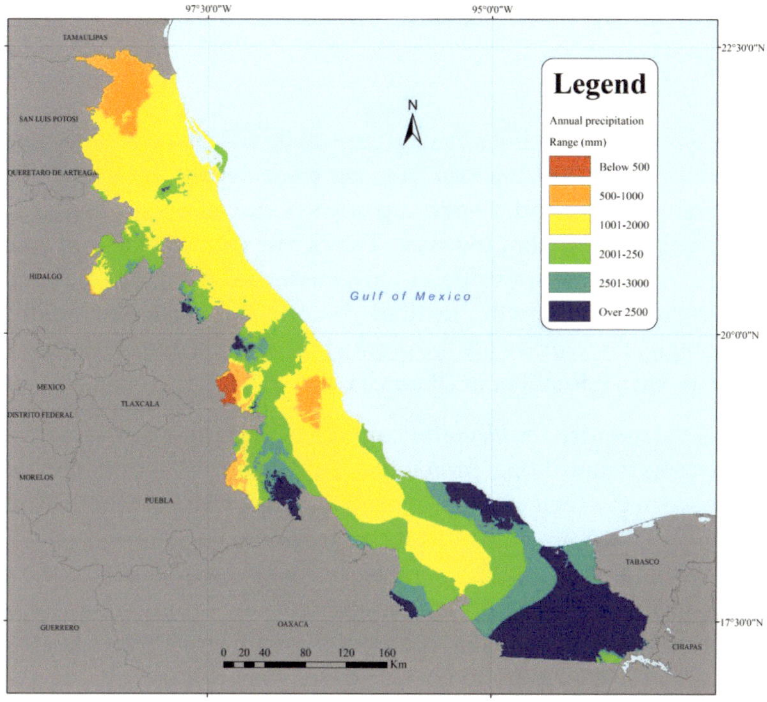

Figure 4. Annual precipitation in Veracruz.

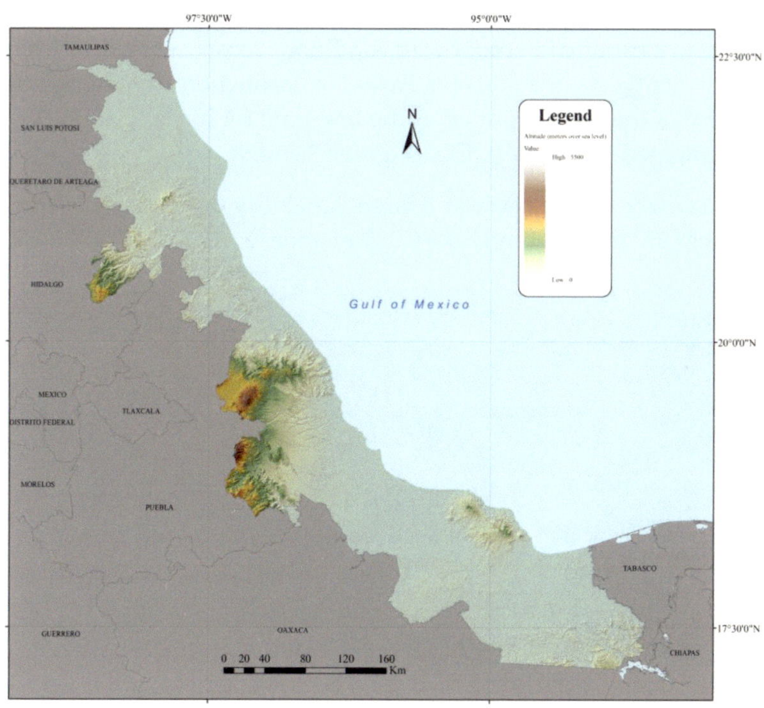

Figure 5. Slopes in Veracruz.

3. Nonconventional by-products and waste biomass from Veracruz for *Pleurotus* cultivation

Pleurotus species represent a well-defined group of Basidiomyceteous fungi of the order Agaricales and family Tricholomataceae. They are characterized by the production of fruit bodies with an eccentric stalk and a wide cap shaped like an oyster shell, with the widest portion of the cap being away from the stalk. They grow over a wide range of temperature and are able to colonize a wide spectrum of unfermented, natural, lignino-cellulosic wastes. Because of their fast mycelial growth rate, they colonize the substrates rapidly; the yield of fruit bodies is also high [23]. Worldwide the production of edible mushroom is widely distributed; however, in Mexico, it is still poor (**Figures 6** and **7**)

The importance on the research for *Pleurotus* cultivation is the alternative to transform a wide range of organic wastes to nutritional protein food with easy access and low cost in Veracruz Mexico[26]. There are many reviews analyze the *Pleurotus* production on different substrates [27–31]. In this study, we assessed the mushroom production on 30 nonconventional agricultural by-products and agro-industrial wastes and some combinations for the cultivation of *Pleurotus*. We used a commercial strain of *Pleurotus ostreatus* supplied by a supplier national laboratory. The wastes biomass was collected from the main crops representative from different regions in Veracruz. The preparation of substrates were completely dried under the sun, chopped, washed, soaking in cold water and sterilized by immersion in hot water (80°C) during 1 h; later when substrate was at room temperature, it was inoculated into bags or plastic container with 2.85 kg of substrate (wet weight) and 0.150 kg of spawn (5%). The incubation was in darkness condition at room temperature, when the primordium was observed the bags were transfer to a production room with conditions recommended by literature [32–34]. We presented a diversity of substrates for *Pleurotus* production from agricultural wastes in Veracruz, and some of these are reported as the first time for its cultivation. The control substrate was old corrugated container (OCC) (**Figures 8** and **9**)

According to the substrate, we observed a large variability on mushrooms production. The biological efficiency (BE) varies from 17.6% (Chilean pine leaves) to 180% (Pineapple fiber and

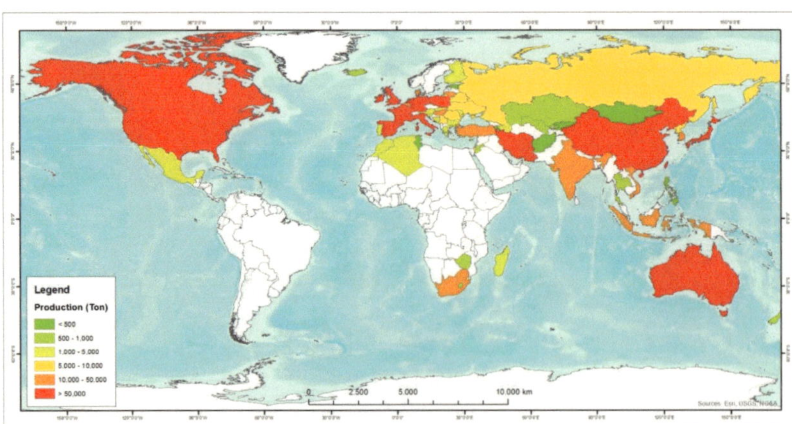

Figure 6. Edible mushroom world production [24].

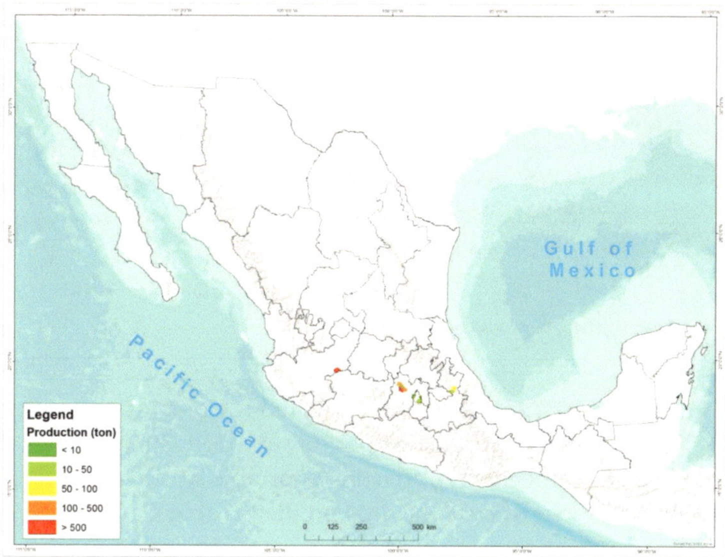

Figure 7. Edible mushroom production in Mexico [25].

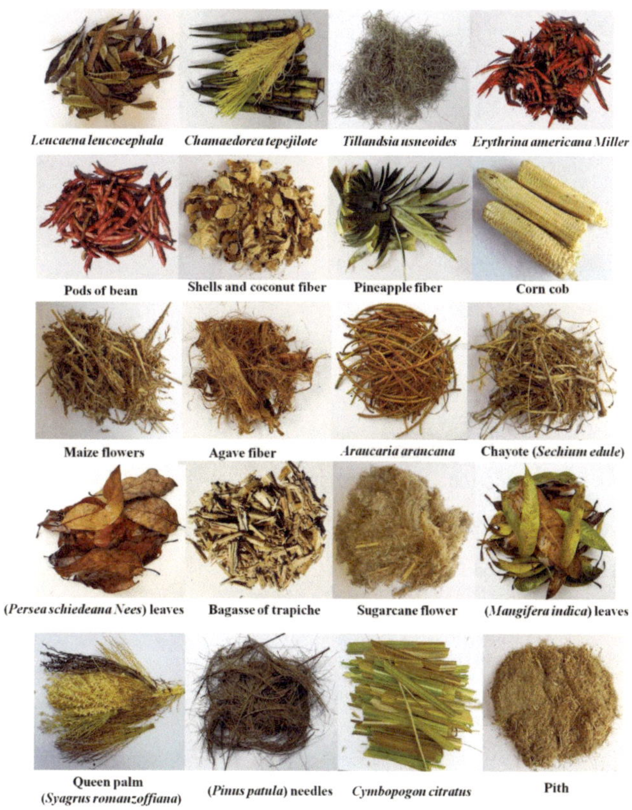

Figure 8. Some wastes as substrates used in the production of *Pleurotus* edible mushroom.

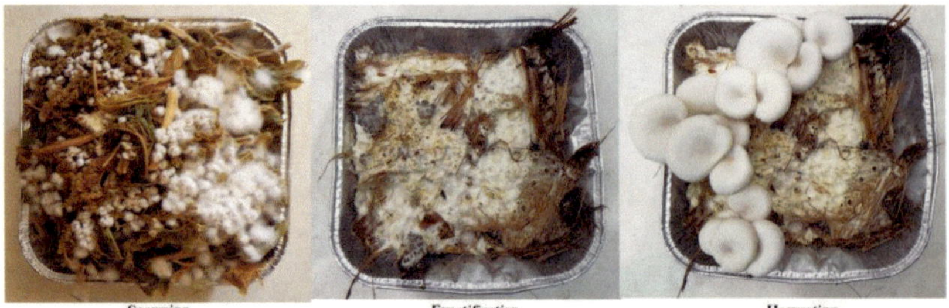

Figure 9. Stages of the edible mushroom production.

pods of guaje), more than 60% of substrates presented a biological efficiency (BE) equal or higher than 50%. Nevertheless, the evaluated period vary from 20 to 96 days (steam of garlic, sugar cane tops, and OCC, respectively), affecting enormously the production rate (PR) from 0.7 to 5.8 and physical characteristics of the final product (**Table 4** and **Figures 10** and **11**).

Substrate	BE(%)	Days of treatment	Production rate	Yield
Chilean pine leaves (*Araucaria araucana*)	17.65	23	0.77	5.61
Tropical wood sawdust	25.00	28	0.89	6.17
(*Pinus patula*) needles	33.33	24	1.39	8.47
Queen palm leaves (*Syagrus romanzoffiana*)	34.62	26	1.33	6.25
Agave fiber	35.00	27	1.30	7.07
(*Mangifera indica*) leaves	35.71	24	1.49	9.62
Sugarcane pith	40.00	23	1.74	8.89
(*Yucca elephantipe*) leaves	45.00	25	1.80	8.65
(*Tillandsia usneoides*) Pascle	46.15	21	2.20	7.59
Cempasuschil (*Tagetes erecta*) stalks	50.00	22	2.27	12.50
Stem of garlic	50.00	20	2.50	10.34
Maize flower	50.00	27	1.85	8.77
Sugar cane flower	55.00	22	2.50	9.91
Lemongrass (*Cymbopogon citratus)*	58.33	26	2.24	8.64
Shells and coconut fiber	59.00	84	0.70	12.29
Corn cob	61.67	28	2.20	15.42
(*Persea schiedeana* Nees) leaves	62.50	23	2.72	8.47
Banana wastes	65.00	25	2.60	12.04
(*Chamaedorea tepejilote* Liebm) leaves	72.22	23	3.14	11.30
Stem of onion	84.62	21	4.03	12.50

Substrate	BE(%)	Days of treatment	Production rate	Yield
Tomato straw	90.00	42	2.14	15.00
Bagasse from trapiche	100.00	26	3.85	13.51
Sugarcane tops	120.00	85	1.41	31.76
Pea pods (*Pisum sativum*)	125.00	25	5.00	24.04
(*Erythrina americana* Miller) flowers	135.0	23	5.87	22.69
Pods of bean	142.5	60	2.38	67.86
Chayote (*Sechium edule*) straw	150.0	40	3.75	16.67
Pineapple fiber	180.0	54	3.33	90.00
Pods of guaje (*L. leucocephala*)	180.0	53	3.40	90.00
Old corrugated container (OCC)	48	96	0.5	12.88

BE: Biological efficiency (fresh mushroom weight/dry substrate weight); Days of treatment: Days from inoculation until the last harvest; Production rate: BE/days of treatment; Yield: fresh mushroom weight/fresh substrate weight.

Table 4. By-products and waste biomass from Veracruz evaluated for *Pleurotus* cultivation.

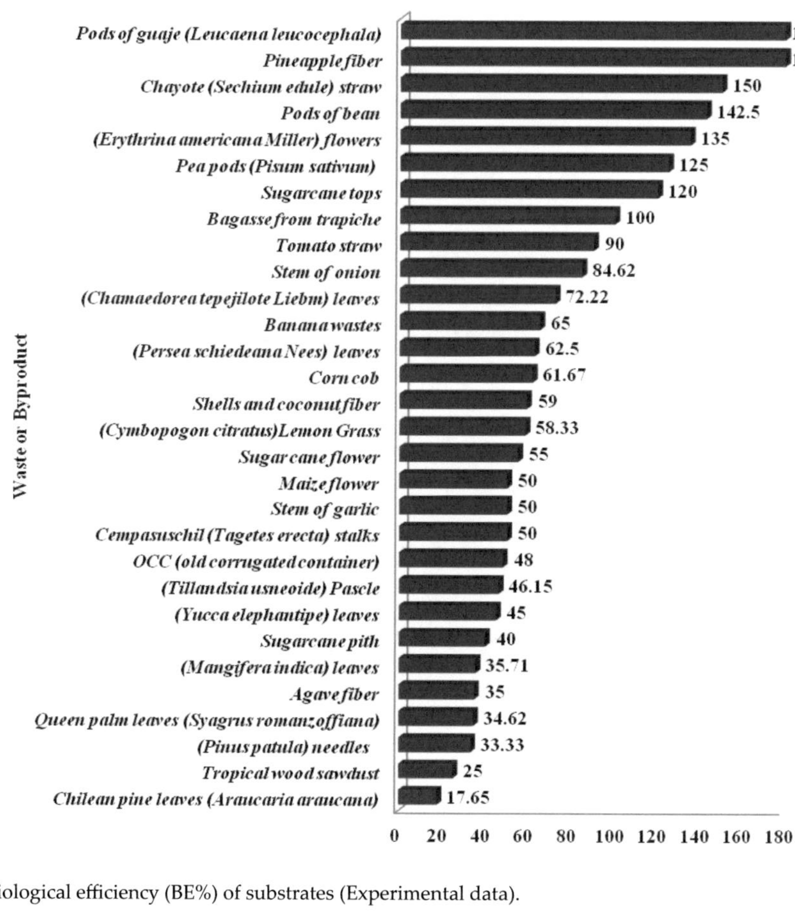

Figure 10. Biological efficiency (BE%) of substrates (Experimental data).

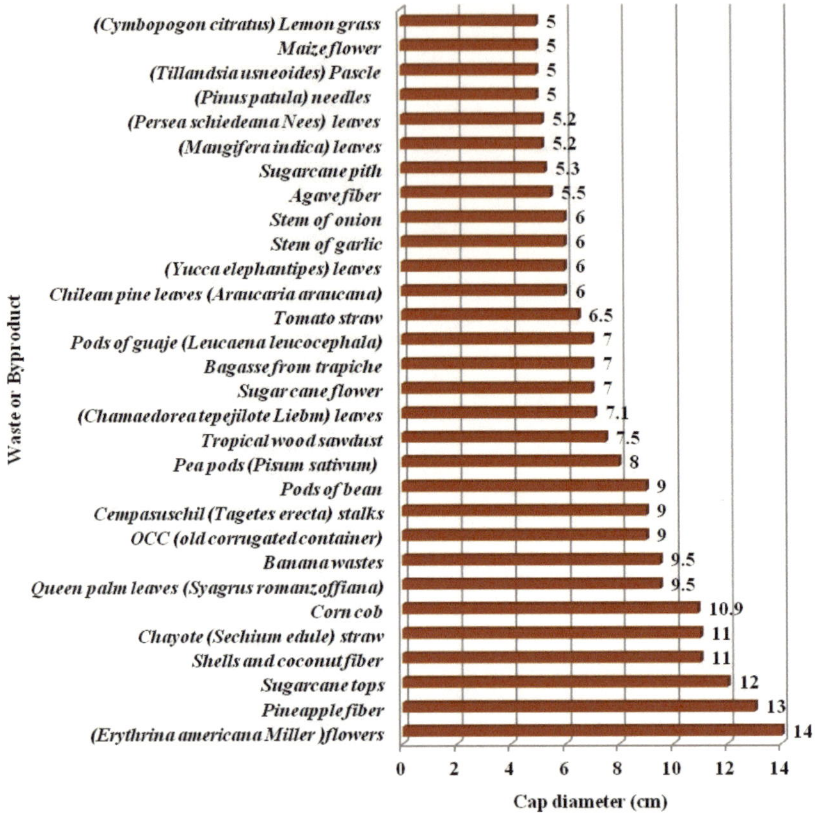

Figure 11. Physical characteristics of produced edible mushroom (Experimental data).

4. Potential of substrates for *Pleurotus* production

The highest BE were observed on the pods of guaje (*Leucaena leucocephala*) and pineapple fiber, both with 180%. Our results are consistent with a recent report of *L. leucocephala* for *P. ostreatus* cultivation, which obtains 157% of BE [35]. They have observed an increase in the nutritional value of sporophores; however, in another study, when *L. leucocephala* is mixed with rice straw [36], the BE has been 65%. In the case of the Pineapple fiber as a substrate, there exist few reports with the objective to obtain fructification of *Pleurotus*. One study [37] reported a BE of 103% with *P. ostreatus*, but on the other hand, however [38, 39] proved that the wastes from this plant stimulate the enzyme of *Pleurotus*, particularly important to obtain by-products. These two substrates represent an excellent choice to obtain a high biomass production for *P. ostreatus* in Veracruz.

Beyond the BE, is important to consider the time to production, because a rapid spawn run would reduce the time noncolonized substrate when it is exposed to competitors such as weed mold and bacteria, this affect the cost production and the risk of contamination. In this study, we consider the production rate (PR) as the division between BE on the days to production. Interestingly was the case of Bean, guaje (*L. leucocephala*) and legume pods mix with a BE of 125% and PR 5.0 and *Erythrina americana* flowers with a BE of 125% and 5.87 of PR, the evaluated period for production was the 25 and 23 days respectively. The first substrate includes the *L. leucocephala,* it is well known to have a high content of Nitrogen [40, 41], and this can help to supply the nutrition of *Pleurotus* to increase the biomasse as mentioned earlier. This is the first report of the *Erythrina americana* flowers, used as a substrate for *Pleurotus* cultivation. This flower is an edible flower from wild plants consumed in some areas in Mexico. In Veracruz, it is consumed during a short time of the season when they are blooming, generally in February, March, and April [42, 43]. Currently, there has not a commercial use of this flower, then to collect and stock this waste during this season could be a strategy for *Pleurotus* cultivation during the year.

On the other hand, the most part of substrates presented a BE higher than 40% and a PR higher than 1 (except the Shells and coconut fiber), these are Sugarcane pith, *Yucca elephantipes* leaves, *Tillandsia usneoides, Tagetes erecta* stalks, Stem of garlic, Sugar cane flower, *Cymbopogon citratus*, Corn cob, *Persea schiedeana* Nees leaves, Banana wastes, *Chamaedorea tepejilote* Liebm, Stem of onion, Tomato straw, Bagasse from trapiche and Sugarcane tops. From those, the

Pith sugar cane [44, 45], Yucca [46], steam of garlic [47, 48], *Cymbopogon citrates* [49], corn cob [30, 35, 47, 49] and Banana wastes [35, 50–52], Onion [53] (stimulation of enzymes of *Pleurotus*), Tomato straw [36, 54], and Pod of beans [49] have been reported, but this is the first report for *Tillandsia usneoides, Tagetes erecta* stalks, Sugar cane flower and *Persea schiedeana* Nees leaves and Chamaedorea tepejilote Liebm leaves as substrates for the *Pleurotus* cultivation. All these substrates represent an economical opportunity, as is reported for *Pleurotus* that a minimum of 40% of BE is economically viable [55]. From these substrates, the banana wastes have been reported as a good option to produces *P. ostreatus* [35, 51, 56], and recently reported that it can increase the lacassa activity of vegetative phase of *P. ostreatus* [57]. Therefore, we consider the banana waste is interesting substrates to develop a standard process for cultivation of *Pleurotus*.

5. Sugar cane by-products

It is particularly important the case of the by-products and wastes of sugar cane. In this study, we use four by-products from this crop: sugarcane pith (40% BE), sugar cane flower (55% BE), bagasse from trapiche (100% BE), and sugarcane tops (120% BE), all of them with an important result for production. Veracruz is the highest producer of sugar cane in Mexico [58], (**Figure 12**) there are big amounts of by-product without a specific use, this should be one of the main substrates for *Pleurotus* production in this state, and this allow to obtain a large amounts of substrate, from short distance and at low price. There are several reports about the use different by-products from sugar cane as substrates for *Pleurotus* cultivation with interesting results for yield [59–63]. This is a no profited by-product with a high potential to transform wastes to nutritional biomass for the economic development.

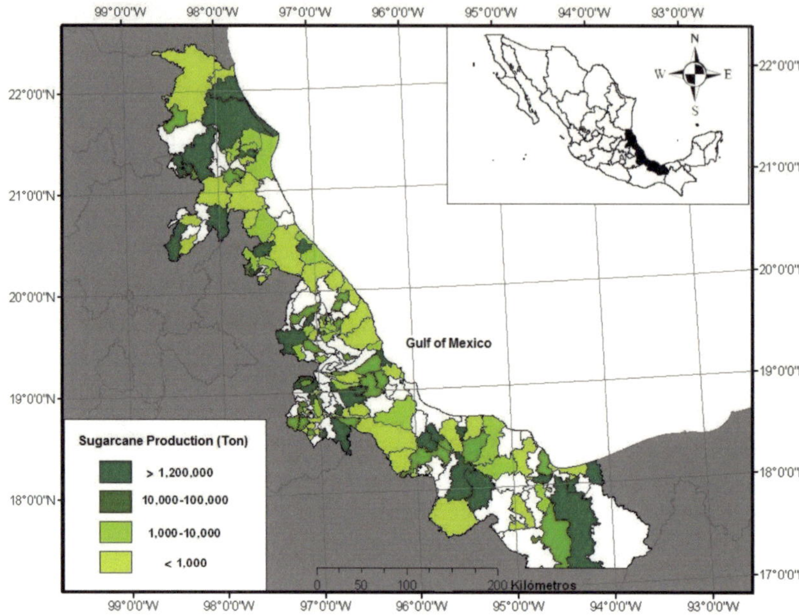

Figure 12. Sugarcane production in Veracruz Mexico [25].

6. Diversity of substrates during the year

In general, the importance of all these substrates here presented is that there are available on different seasons during the year; the price is usually is very low or even for free, thus this is the opportunity to develop a plan production to use specific substrates during the year. Verma [64]analyze an economic model for a farm for *Pleurotus* cultivation based on coffee pulp on México, which could be improved by the adding the sugar cane wastes and other agricultural by-products mentioned earlier.

The high prices of raw materials used as substrates for the mushroom cultivation could increase the cost of commercial production (*P. ostreatus*). In Mexico, the wheat straw and barley straw are the most common substrates for cultivation of *Pleurotus* [64, 65], while the 77% of this crops is produced in Sonora y Baja California, and the by-products come from the northern of Mexico [66, 67], and their availability vary during the year and as the same as the price. Nowadays, there is some sugar mill research for different use for bagasse, which subsequently may avoid ecological problems as pests and lixiviation produced by the accumulation of big amounts of wastes, this process is also necessary to avoid ecological problems as pests and lixiviation produced by the accumulation of big amounts of wastes. It is an opportunity to develop a plan including the mix of the most important wastes from this region for plan *Pleurotus* cultivation and other tropical edible mushrooms.

Into the plan, is important consider some characteristics of the substrates, according to [59], the materials used for substrates of *Pleurotus* should count with positives properties as (1) availability of amount and time, (2) knowledge of their physic-chemical content, (3) stability on their physic-chemical content, (4) advantage on cost, (5) easy location on short distance and (6) easy to transport. For some substrates here presented, is important to consider the points two and 3, being that is the first reports for *Pleurotus* cultivation, and then more studies are necessaries to guaranty the safety on the long term.

7. Composting edible mushroom wastes

In order to reuse the spent substrates from the production of edible mushroom *P. ostreatus*, compost production was exclusively used grass pruning residue, unlike other conventional inputs such as sugar industry byproducts (filter mud, bagasse, and vinasses) or cattle manures used in composting. The materials used once weighed were placed on a layered plastic to form a static pile with a rectangular pyramidal structure: The layers were formed with fresh grass, edible mushroom waste and covered with fresh grass until exhausted the substrates to complement 40 kg of fresh grass and 40 kg of edible mushroom waste.

The grass was used as the initiator of the microbial growth process to reach the thermophilic phase temperatures indispensable in composting. The initial moisture content of the grass was 82% and for waste edible mushroom 25%, so water was added so that the moisture of the pile formed was 75%. During the experimental process, the pile moisture remained in the 65–70% range [68, 69]. The figure details the dimensions of the pile made with the previously mentioned substrates (**Figure 13**).

The composting process can be considered as an effective method to transform organic matter from edible mushroom production into a potentially safe, stable and harmless material that can be used as a soil improver, organic fertilizer, or manure substitute in crop soils (Chen et al., 2014). In the case of composting edible mushroom production wastes, the maximum temperature reached was 50°C, and this is attributed to the fact that the substrates were partially degraded by the mushroom growth process, thereby reducing a number of nutrients available in it.

At the same time, vermicompost and bocashi were obtained with edible mushroom production wastes which in turn was mixed with by-products from sugar and livestock industry by the conventional methods according to [70, 71] and evaluated for their nutritional value according to Mexican standard [72] (**Table 5**).

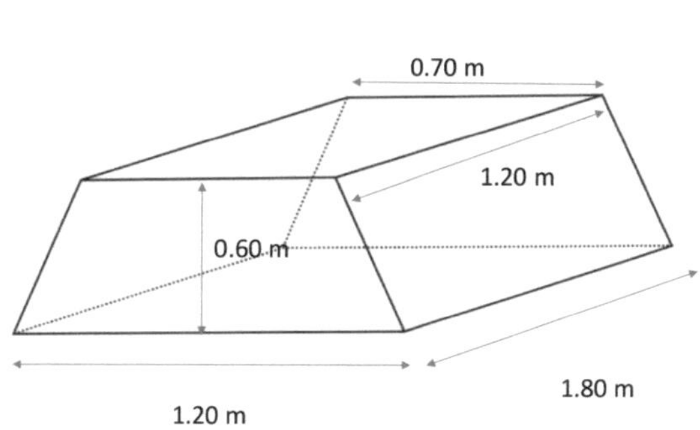

Figure 13. Composting static pile.

Variable	Unit	Compost*	Vermicompost**	Bocashi***
Humidity	%	49.31	48.98	44.77
pH		7.54	7.27	8.99
Soil electrical conductivity	dsm^{-1}	4.080	1.076	3.77
Ash	%	29.9	50.01	50.52
Organic matter	%	70.02	49.99	49.48
Total carbon C	%	40.615	28.996	28.701
Total nitrogen N	%	1.51	0.51	0.9
C/N		26.90	56.85	31.89
Ca	%	3.339	4.849	25.233
Mg	%	0.576	0.610	0.703
Na	%	0.102	0.029	0.619
K_2O	%	1.65	0.156	1.216
P_2O_5	%	1.258	3.711	1.071
Fe	%	0.0283	0.7310	0.1515
Cu	%	0.0018	0.0051	0.0024
Zn	%	0.0170	0.0324	0.0147
Mn	%	0.0377	0.1950	0.0715

*Compost from waste from cultivation of oyster mushrooms and grass pruning residue.
**Filter mud vermicompost and waste from cultivation of oyster mushrooms.
***Filter mud, cattle manure, ashes, lime and waste from cultivation of oyster mushrooms.

Table 5. Chemical composition of organic manures from edible mushroom wastes.

The content of organic matter in the compost at the end of the process indicates that once it is applied in the soil, it will continue with the degradation process, the nitrogen present indicates that during the cultivation of mushroom there was a significant consumption of the same. However, unlike the vermicompost and bocashi, the edible mushroom waste compost, obtained by the pile method, presents a neutral pH, higher carbon content, organic matter, and potassium and lower C/N ratio.

The final mass obtained was 10 kg, with a yield of 12.5%. (**Figure 14**) Composting of edible mushroom production wastes can be a technology of production of foods such as vegetables and organic manures at the same time; besides to manage the organic fraction of biomass in developing countries due to the simplicity and speed of implementation. This technology implies low costs in its application, compared to other processes, which require a strong investment for the installation, management, and operation of the necessary equipment [73].

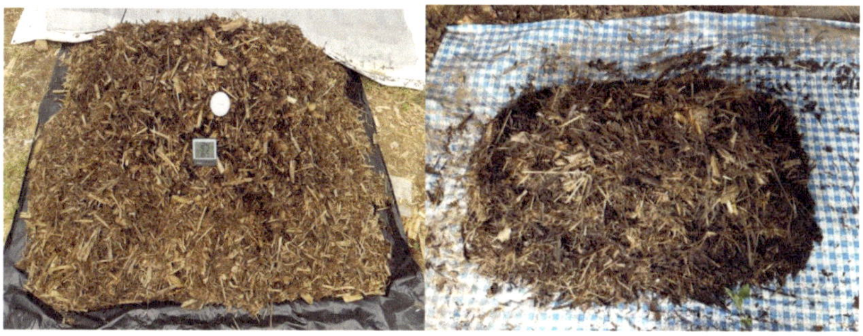

Figure 14. Compost of edible mushroom wastes (Pile method) [Experimental data].

8. Conclusions

We presented the results of an evaluation of 30 nonconventional by-products and waste biomass and mixtures useful for *Pleurotus* cultivation. From those substrates, 23 of them showed a BE higher than 40%, that represent a minimum to be considered as economically viable. Seven of these substrates are reported as the first time for *Pleurotus* cultivation; however, more studies are necessary to know the physic-chemical properties, characteristics, and the stability for the production of *Pleurotus* in the long term.

The sugar cane represents one of the most important resources, being that Veracruz is the main producer of this crop in Mexico, and this opens the possibility to obtain a large amount of substrate for *Pleurotus*, in a short distance and at a low price.

It is important to develop a plan including the substrates analyzed to propose an economic plan for *Pleurotus* production and composting of waste substrates based on the local crops according to the season and region. It will allow to minimize cost for acquisition and transport of substrates for *Pleurotus* production as well as to avoid the accumulation of large amount of substrates that can produce pollution by pests and lixiviation improving with this closed cycle technology the competitiveness and sustainability of food production.

Author details

Noé Aguilar-Rivera[1]*, Régulo Carlos Llarena-Hernández[1], Christian Michel-Cuello[2], Martin Roberto Gámez-Pastrana[1] and Teresita de Jesús Debernardi-Vazquez[1]

*Address all correspondence to: naguilar@uv.mx

1 Faculty of Biological and Agricultural Sciences, University of Veracruz, Veracruz, Mexico

2 Multidisciplinary Academic Unit Central Zone, Autonomous University of San Luis Potosi, San Luis Potosi, Mexico

References

[1] FAO. The State of Food Insecurity in the World 2015. Fulfillment of the international objectives for 2015 in relation to hunger: Balance of the uneven progress. Food and Agriculture Organization of the United Nations. Roma. 2015. Available from: http://www.fao.org/hunger/en/[Accessed: March 1, 2017]

[2] Soria SG, Palacio MVH. The actual scene of de feeding in México. [Internet]. 2014. Available from: http://www.academia.edu/12893339/The_Actual_Scene_of_the_Feeding_in_Mexico. [Accessed: February 11, 2017]

[3] Urquía-Fernández N. Food security in México. Mexican Public Health. 2014;**56**(1):592-598

[4] Fausto GJ, Valdez L RM, Aldrete RMG, López ZMC. Social historical antecedents of obesity in Mexico. Health Research. 2006;**VIII**(2):91-94

[5] Aguirre P. Socioanthropological Aspects of Obesity in Poverty. In: Peña M, Bacallao J. (Comp.). Obesity and Poverty: A New Public Health Challenge. Scientific Publication. Pan-American Health Organization; 2000;**576**:11-22

[6] FAO. Program to Fight Hunger. Dual Component Approach to Hunger Reduction: Priorities for Action at the National and International Levels. [Internet]. 2003a Available from: ftp://ftp.fao.org/docrep/fao/006/j0563e/j0563e00.pdf [Accessed: March 12, 2017]

[7] Román S, Ojeda-Granados C, Panduro A. Genetics and food evolution of the population in Mexico. Journal of Endocrinology and Nutrition. 2013;**21**(1):42-45

[8] INEGI. Sociodemographic Indicators of Mexico (1930-2000). [Internet]. 2001 Available from: http://internet.contenidos.inegi.org.mx/contenidos/productos/prod_serv/contenidos/espanol/bvinegi/productos/integracion/pais/mexvista/mexatg.pdf [Accessed: March 12, 2017]

[9] FAO. Nutrition Profiles by Country—Mexico. Food and Agriculture Organization of the United Nations. Roma. [Internet]. 2003. Available from: http://www.fao.org/countryprofiles/index/en/?iso3=MEX [Accessed: March 10, 2017]

[10] Ibarra-Sánchez LS, Viveros-Ibarra LS, González-Bernal V, Hernández-Guerrero F. Mexico Food Transition. Razón y Palabra. 2017;**20**:166-182

[11] Rivera JA, Pedraza LS, Aburto TC, Batis C, Sánchez-Pimienta TG, González de Cosío T, López-Olmedo N, Pedroza-Tobías A. Overview of the dietary intakes of the Mexican population: Results from the National Health and Nutrition Survey 2012. The Journal of nutrition. 2016;**146**(9):1851S-1855S.

[12] Gómez F. Malnutrition. Public Health of Mexico. 2003;45(4):s576-s582

[13] Shamah LT, Amaya CMA, Cuevas NL. Malnutrition and Obesity: Double Burden in Mexico. [Internet]. 2015. Available from: http://www.revista.unam.mx/vol.16/num5/art34/[Accessed: March 12, 2017]

[14] Hernández-Ávila M, Gutiérrez JP, Reynoso-Noverón N. Diabetes mellitus in Mexico. The state of the epidemic. Public Health in Mexico. 2013;**55**(2):s129-s136

[15] Estrada ME, Aguirre RJR. Feeding the old Mexicans. In: Sánchez VA. (Comp.). The Destruction of Indias and its Renewable Resources 1492-1992. México: UACh, Chapingo; 1999, pp. 141-152

[16] Sarukhán J, Koleff P, Carabias J, Soberón J, Dirzo R, Llorente-Bousquets J, Halfter G, González R, March I, Mohar A, Anta S, de la Maza J. Natural capital of Mexico. Synthesis: current knowledge, evaluation and sustainability perspectives. National Commission for the Knowledge and Use of Biodiversity. México. 2009. CONABIO, p. 100

[17] de Moraes Sá JC, Lal R, Cerri CC, Lorenz K, Hungria M, de Faccio CarvalhoPC. Low-carbon agriculture in South America to mitigate global climate change and advance food security. Environment International. 2017;**98**:102-112

[18] CEC. Burning Agricultural Waste: A Source of Dioxins.Montreal, Canada: Commission for Environmental Cooperation; 2014. p. 6

[19] Koopmans A, Koppejan J. Agricultural and Forest Residues-generation, Utilization and Availability. 1997. Available from: http://www.fao.org/docrep/006/AD576E/ad576e00.pdf [Accessed: March 12, 2017]

[20] FAO. The State of Food and Agriculture. Climate Change, Agriculture and Food Security. Roma, Italia. 2016. Available from: http://www.fao.org/3/a-i6030e.pdf [Accessed: February 28, 2017]

[21] Olson DM, Dinerstein E, Wikramanayake ED, Burgess ND, Powell GVN, Underwood EC, D'Amico JA, Itoua I., Strand HE, Morrison JC, Loucks CJ, Allnutt TF, Ricketts TH, Kura Y, Lamoreux JF, Wettengel WW, Hedao P, Kassem KR. Terrestrial ecoregions of the world: A new map of life on Earth: A new global map of terrestrial ecoregions provides an innovative tool for conserving biodiversity. BioScience, 2001;**51**(11): 933-938

[22] Galindo-Tovar ME, Aguilar-Rivera N, Gámez Pastrana MR, Leyva-Ovalle OR, Murguía-González J. 2015. Advances in Research on Tropical Horticulture in Sugarcane Crop Fields from Veracruz, Mexico. Ebook Series: Agricultural Research Updates. Available from https://www.novapublishers.com/catalog/product_info.php?products_id=55152&osCsid=a747dcf19b64b3288b2ad5b14afe591f.[Accessed: March 10, 2017]

[23] Rajarathnam S, Bano Z. *Pleurotus* mushrooms. 1A. Morphology, life cycle, taxonomy, breeding, and cultivation. CRC Critical Reviews in Food Science and Nutrition. 1987;**26**(2):157-223

[24] FAOSTAT. 2016. Available from http://www.fao.org/faostat/en/#data/QC [Accessed: March 12, 2017]

[25] SIAP. Anuario Estadístico de la Producción Agrícola. Available from: http://infosiap.siap.gob.mx/aagricola_siap_gb/icultivo/index.jsp>[Accessed: January 25, 2017]

[26] Chang S. World production of cultivated edible and medicinal mushrooms in 1997 with emphasis on Lentinus edodes (Berk.) Sing, in China. International Journal of Medicinal Mushrooms. 1999;**1**:291-300

[27] Pineda-Insuasti JA, Ramos-Sánchez LB, Soto-Arroyave CP, Ciudadela AA EP, Victoria L, Ecuador I. Producción de *Pleurotus ostreatus* por fermentación en estado sólido: una revisión. ICIDCA. Sobre los Derivados de la Caña de Azúcar. 2014;**48**:13-23

[28] Josephine RM. A review on oyster mushroom (*Pleurotus* spp). International Journal of Advanced Biotechnology and Research. 2014;**1**:7-9

[29] Díaz C, Carvajal R. Biological efficiency of *Pleurotus ostreatus* cultivated in palm fiber oil. @Limentech Ciência e Tecnologia de Alimentos. 2014;**12**:63-70

[30] De Carvalho CSM, Sales-Campos C, De Andrade MCN. Mushrooms of the *Pleurotus* genus: A review of cultivation techniques. Interciencia. 2010;**35**:177-182

[31] Rajarathnam S, Bano Z, Miles PG. *Pleurotus* mushrooms. Part I A. morphology, life cycle, taxonomy, breeding, and cultivation. CRC Critical Reviews in Food Science and Nutrition. 1987;**26**:157-223

[32] Gaitán-Hernández, R., Salmones, D., Pérez Merlo, R., & Mata, G. (2006). *Manual práctico del cultivo de setas: aislamiento, siembra y producción (1a edición). Xalapa: Instituto de Ecología* A.C. http://doi.org/970-709-042-1

[33] Oei P. Small-scale Mushroom Cultivation. Oyster, Shiitake and Wood Ear Mushrooms. Wageningen, The Netherlands: Digigrafi; 2005

[34] Abeer H, Alqarawi AA, Al-Huqail AA, Abd Allah E. Biochemical composition of *Pleurotus ostreatus* (Jacq.) P. Kumm. grown on sawdust of *Leucaena leucocephala* (Lam.) De Wit. Pakistan Journal of Botany. 2013;**45**:1197-1201

[35] Mamiro D, Mamiro P. Yield and mushroom size of *Pleurotus ostreatus* grown on rice straw basal substrate mix*ed and supplemented* with various crop residues. Journal of Animal and Plant Sciences. 2011;**10**:1211-1218

[36] Lazo-Lemun G. Determinación de la eficiencia del rastrojo de tomate (Lycopersicon sculentum Miller) y la corona del fruto de piña (*Ananas comosus* L – Merril) y sus mezclas en el cultivo de cela cepa esc 0110 de *Pleurotus* (*Pleurotus ostreatus*). Thesis Universidad de San Carlos de Guatemala, Guatemala. 2001. Available from: http://biblioteca.usac.edu.gt/tesis/01/01_1992.pdf [Accessed: March 12, 2017]

[37] Teixeira Souza RA, et al. Nutritional composition of bioproducts generated from semi-solid fermentation of pineapple peel by edible mushrooms. African Journal of Biotechnology. 2016;**15**:451-457

[38] Rashad, M. M., Abdou, H. M., & Mahmoud, A. E. (2016). Production of some bioactive materials by Pleurotus ostreatus from pineapple residues and rice straw via solid state fermentation. *Research Journal of Pharmaceutical, Biological and Chemical Sciences*, 7(**5**):2730-2736.

[39] Garcia GW, Ferguson TU, Neckles FA, Archibald KAE. The nutritive value and forage productivity of *Leucaena leucocephala*. Animal Feed Science and Technology. 1996;**60**:29-41

[40] Kang B, Sipkens L, Wilson G, Nangju D. Leucaena (*Leucaena leucocephala* (Lam) de Wit) prunings as nitrogen source for maize (*Zea mays* L.). Fertility Research and Practice. 1981;**2**:279-287

[41] Sotelo A, López-García S, Basurto-Peña F. Content of nutrient and antinutrient in edible flowers of wild plants in Mexico. Plant Foods for Human Nutrition. 2007;**62**:133-138

[42] García-Mateos R, Soto-Hernández M, Vibrans H. Erythrina Americana Miller ('Colorín'; Fabaceae), a versatile resource from Mexico: A Review. Economic Botany. 2001;**55**:391-400

[43] Ragunathan R, Gurusamy R, Palaniswamy M, Swaminathan K. Cultivation of *Pleurotus* spp. On various agro-residues. Food Chemistry. 1996;**55**:139-144

[44] Aguilar-Rivera N, Castillo-Moran A, Rodriguez-Lagunez DA, Murguia-González J. In: Andres S, Baumann N, (editors). Mushrooms: Types, Properties and Nutrition. Nova Science Publisers; New York USA. 2012;77-104

[45] Flores Ramírez G. Aprovechamiento del bagazo residual de Yucca spp como sustrato para la producción de *Pleurotus* spp. Thesis Instituto Politécnico Nacional, México. 2012. Available from: http://hdl.handle.net/123456789/19825 [Accessed: March 10, 2017]

[46] Sang-Sun L, Soon-Keun K, Tae-Soo L, Min-Woong L. Cultivation of oyster mushroom using the garlic peel as an agricultural by-product. Korean Journal of Medical Mycology. 1997;**25**:268-275

[47] Lechner, Bernardo E. and Santiago Monaldi. 2011. "Utilization of Garlic and Maize Wastes Supplemented with Olive Mill Waste Water for Pleurotus Ostreatus Cultivation." *Revista mexicana de micología* **34**:17-22.

[48] Silva SO, da Costa SMG, Clemente E. Chemical composition of *Pleurotus pulmonarius* (Fr.) Quél., substrates and residue after cultivation. Brazilian Archives of Biology and Technology. 2002;45:531-535

[49] Alejo-Cristobal P, Martínez-García R, León-Avedaño H. Eficiencia biológica de *Pleurotus* spp nativo de Oaxaca (ITAO-27) sobre rastrojos de frijol, maíz y olote. Revista Mexicana de Agroecosistemas. 2015;**2**:99-106

[50] Obodai M, Cleland-Okine J, Vowotor KA. Comparative study on the growth and yield of *Pleurotus ostreatus* mushroom on different lignocellulosic by-products. Journal of Industrial Microbiology and Biotechnology. 2003;**30**:146-149

[51] Bonatti M, Karnopp P, Soares HM, Furlan SA. Estudo da composição de cogumelos das espécies *Pleurotus ostreatus* e *Pleurotus sajor-caju* cultivados em palha de bananeira. Rev. Saúde e Ambient. 2008;**4**:31-35

[52] Girmay Z, Gorems W, Birhanu G, Zewdie S. Growth and yield performance of *Pleurotus ostreatus* (Jacq. Fr.) Kumm (oyster mushroom) on different substrates. AMB Express. 2016;**6**:87

[53] Shankar Patel S, Birwal P, Singham P, Sushant Sukhdev B, Kumar Nema P. Effect of pre-treatment by onion extract on quality of dried oyster mushrom (*Pleurotus sajor-caju*). Environment and Ecology. 2016;**34**:2591-2595

[54] Ananbeh, K. and A. Almomany. 2008. "Production of Oyster Mushroom (Pleurotus Ostreatus) on Tomato Tuff Agrowaste." *Dirasat, Agricultural Sciences* **35**:133-38.

[55] Rios MM, del P, Hoyos JL, Mosquera SA. Evaluation of parameters production of the seed *Pleurotus ostreatus* spreed in different cuture media. Biotecnología en el Sector Agropecuario y Agroindustrial. 2010;**8**:86-94

[56] Obodai M, Cleland-Okine J, Vowotor KA. Comparative study on the growth and yield of *Pleurotus ostreatus* mushroom on different lignocellulosic by-products. Journal of Industrial Microbiology and Biotechnology. 2003;**30**:146-149

[57] Kurt S, Buyukalaca S. Yield performances and changes in enzyme activities of *Pleurotus* spp. (*P. ostreatus* and *P. sajor-caju*) cultivated on different agricultural wastes. Bioresource Technology. 2010;**101**:3164-3169

[58] Senties-Herrera HE, Gómez-Merini FC, Valdéz-Balero A, Silva-Rojas HV, Trejo-Téllez LI. The agro-industrial sugarcane system in Mexico: Current status, challenges and opportunities. Journal of Agricultural Science. 2014;**6**:26-54

[59] Muez-Orobia, M. A., & Pardo-Nuñez, J. (2001). La preparación del substrato. In J. E. Sanchez & D. J. Royse (Eds.), *La biología y el cultivo de Pleurotus spp*. México: El Colegio de la Frontera Sur. 2001;157-186.

[60] Ahmed S. Performance of Different Substrates on the Growth and Yield of Oyster Mushroom *Pleurotus sajor-caju* (Fr.) Sing. Salna, Gazipur, Bangladesh: Institute of post-graduate studies in Agriculture; 1998

[61] Carrera, M., D. Morales, and MP Sobal. 1990. "Cultivo de Pleurotus Ostreatus Sobre Bagazo de Caña Enriquecido Con Pulpa de Café O Paja de Cebada." *Micología Neotropical Aplicada* **3**:49-52.

[62] Soto-Velasco, C. L., L. Guzman-Dávalos, and C. Téllez. 1991. "Substrates for Cultivation Pleurotus in Mexico, II. Sugar Cane Bagasse and Corn Stover." *Mushroom Journal of Tropics* 11:34-37.

[63] Aguilar-Rivera, N. and J. De Jesús-Merales. 2010. "Edible Mushroom Pleurotus Ostreatus Production on Cellulosic Biomass of Sugar Cane." *Sugar Tech* **12**(2):176-78.

[64] Verma, R. N. (2001). Aspectos económicos de la producción de Pleurotus spp. In J. E. Sanchez & D. J. Royse (Eds.), *La biología y el cultivo de Pleurotus spp*. México: El Colegio de la Frontera Sur. 2001;273-290.

[65] Gaytan-Hernández, R. (2007). Transferencia de tecnología de cultivo de Pleurotus spp como alternativa de beneficio social y económico en el estado de Veracruz. In J. E. Sanchez, D. Martínez-Carrera, G. Mata, & H. Leal-Lara (Eds.), *El cultivo de setas Pleurotus spp. en México*. México: El Colegio de la Frontera Sur. 2007;101-112.

[66] Castañeda-de-Leon VT, Leal-Lara H. In: Sanchez JE, Martínez-Carrera D, Mata G, Leal-Lara H, (editors). El cultivo de setas *Pleurotus* spp. en México. El Colegio de la Frontera Sur. 2007. pp. 81-90

[67] FIRA. Panorama Agroalimentario de trigo. (2014). Available from: https://www.gob.mx/cms/uploads/attachment/file/61954/Panorama_Agroalimentario_Trigo_2015.pdf [Accessed: January 31, 2017]

[68] Robles Martinez F, Gerardo Nieto O, Pina Guzman AB, Montiel Frausto L, Colomer Mendoza FJ, Orozco Alvarez C. Obtaining alternate fuels from the biodrying of fruits and vegetables. Revista Internacional de Contaminacion Ambiental. 2013;**29**:79-88

[69] Yaoning C, et al. Nitrite reductase genes as functional markers to investigate diversity of denitrifying bacteria during agricultural waste composting. Applied Microbiology and Biotechnology. 2014;**98**(9):4233-4243

[70] Saranraj P, Stella D. Composting of sugar mill wastes: A review. World Applied Sciences Journal. 2014;**31**(12):2029-2044

[71] Saldaña María I, et al. The influence of organic fertilizers on the chemical properties of soil and the production of Alpinia purpurata. Ciencia e Investigación Agraria. 2014;**41**(2):215-224

[72] NOM-021-SEMARNAT, 2000. México. Available from: http://biblioteca.semarnat.gob.mx/janium/Documentos/Ciga/libros2009/DO2280n.pdf [Accessed: January 31, 2017]

[73] Jara-Samaniego J, et al. Composting as sustainable strategy for municipal solid waste management in the Chimborazo Region, Ecuador: Suitability of the obtained composts for seedling production. Journal of Cleaner Production. 2017;**141**:1349-1358

Chapter 2

Nutritional Value, Food Ingredients, Chemical and Species Composition of Edible Insects in China

Weiping Yin, Junna Liu, Huaqing Liu and Biyu Lv

Additional information is available at the end of the chapter

http://dx.doi.org/10.5772/intechopen.70085

Abstract

As the prevailing food cultures, edible insects could be dated back to ancient China. It is becoming clear that insect resource can be mass-produced in sustainable development food utilization. China could introduce insects into modern western diets. It is a precious resource considering the nutritional value, food ingredients and chemical composition of species. Meanwhile, the edible insects' consumption also led to a challenge of addressing food security, nature conservation and the erosion of traditional food culture. We summarized the resourceful edible insects containing the nutrition substance, such as essential proteins, amino acids, fatty acids, carbohydrates, vitamins, mineral elements and other functional ingredients with the insect secondary metabolite, including the flavonoids, alkaloids, polysaccharides, hormones and phospholipids, which have high economic value for development and utilization.Based on the history, custom, plasmid resource, production and status of edible insects in China at present, it has been proven that the development of insects food well matches the need for human health in China.

Keywords: edible insects, sustainable resource food, nutritional value, active ingredients with the insect secondary metabolite, chemical composition, Chinese insect diet

1. Introduction

1.1. Discussing edible insect's resources value

Insects are most species-rich resources and one of the largest biological groups in earth organisms. With its wide resources, speed of reproductive growth, lower feed cost, edible insects are an important food source which has immense potential of the development and utilization. The edible insects contain the reasonable structure of nutrition, high nutritional value and numerous functional factors. In addition, the current research results provided a reliable technical support for the large-scale production and processing of edible insects in factory production. Therefore, edible insects have a very broad prospect as food sustainable resources development in future.

1.2. Human insectivorous history and customs

Though insects actually could be used as the secured and sustainable food, people naturally disgust the pests and have the feeling that they will bring heavy disaster to human beings. Most people will feel incredible even nauseous if they see a person eating insects. Of course, for those who have tasted barbecue locusts' good things, situation is not so strange. However, in fact, if the history window of mankind is opened, we will find that insects occupy a prominent position in the human diet since the ancient times. Countries around the world in many parts of many societies used to have the habit of eating insects [1]. According to the literature reported, the Middle East people ate desert locusts before the 8th century BC [2–4]; in ancient Rome, people liked to eat a kind of larva of *Cossus cossus orientalis* Gaede and prompted to insect body hypertrophy using the flour [2]. In addition, the ancient Hebrew people hunt locusts for eating, American Indians ate the locust pest braised in soy sauce, Australian people loved eating cutworms, Africans fed termites, South Americans hobby to ten red ants and the Spanish make worm bean sauce from ant eggs, etc. Even up to now, the locusts have been accustomed to dry and grinded them into flour, baking cookies or bread sharing in Europe. African residents of some place even take ants, termites that make the taste delicious. In southern California and Mexico, American Indians collect a vast number of backstroke eggs in the water for consumption and sale. The blue butterfly Larvae are more popular in the United States and Mexico; their price is expensive and they are the famous rare dishes in restaurant or hotel. The edible insects also have a long history in China. In the third century BC, the emperor's banquet was made up of cicadas and bees such as Yiluan (ant egg sauce), Huangchong (locust pest), Mifeng (bee) and Chan (cicada), which had been listed as the emperor's own meat food and the banquet delicacy of aristocratic gatherings. There is a traditional Chinese famous specialty named Eight Jane Cakes from insects fly maggots [3]. In conclusion, the edible insects are rich whether they are in quantity or on people's table, including their nutritive value and medicinal value. These insects will play a very important role as a sustainable food resources development in the future. Edible insects from all over the world are listed in **Table 1**.

Insect	The state of edible insect	Countries and regions	Insect	The state of edible insect	Countries and regions
Locust (including migratory locust etc.)	Adult	China, Japan, Vietnam, Thailand, India	Leafspinner ant	Adult	Burma, Thailand, Australia, Malaysia
		Indonesia, America Mexico, Africa, Australia	Honey ant	Adult, larva	America, South Ameirica, Mexico
Gryllotalpa	Adult	China, Japan, Vietnam, Thailand	Cicada	Larva	China, Japan, South America, France

Insect	The state of edible insect	Countries and regions	Insect	The state of edible insect	Countries and regions
Crickets (including oil. Reed big crickets, etc.)	Adult, larva	Indonesia, Japan, Thailand, Malaysia, America, Africa	Mayfly	Adult	China, Japan, Vietnam, America, Africa
Hawkmoth worm	Larva, pupa	China, Africa, Japan, America	Dragonfly	Adult, larva	Japan, Thailand, Indonesia, Africa
Slug moth	Pupa	China, Japan	Caddis worm	Adult, larva	Japan
Pine caterpillars	Larva, pupa	Japan, North Korea, Australia, Japan, America, Mexico	Osmanthus cicada (including negative ducking Rho)	Adult	China, Japan, Malaysia, Burma, Vietnam, Thailand, Australia, America
Noctuidae (cutworm)	Larva, pupa	Japan, South Africa, Australia, China	Aspongopus	Adult	China, India, Mexico, Africa
Locates the moth (including cordyceps sinensis)	Pupa	Japan, Thailand, Italy, Australia	Stinkbug	Adult	China, India, Africa
			Maggots	Larva	China, Japan, Africa, America
Carpenter moth	Larva	Africa			Mexico
Lappet moth	Larva	Africa	Ephydrid	Adult, larva	China, Japan, Thailand
White moth	Larva, pupa	Japan	Cockchafer	Adult, larva	Africa, America, France
Grape wing bug moths	Larva	Japan			Germany
Brahmaeidae	Larva	China, Japan			China, Japan, Thailand
Domain moth (backpack worm)	Adult, larva	Japan, Mexico	Longicorn beetle	Larva	Indonesia, Ceylon, Vietnam, Australia, Africa
Snout moth (including corn moth two group three moth, etc.)	Larva, pupa	Japan	Weevil	Larva	Indonesia, India, Thailand, Burma, Vietnam, Africa, China, Japan, America
Papilionid	Larva	Japan			Thailand, Africa
Skipper	Larva	Mexico	Buprestid beetle	Larva	China

Insect	The state of edible insect	Countries and regions	Insect	The state of edible insect	Countries and regions
Termite	Queen ant, Adult	China, Indonesia, India	Yellow mealworm	Pupa	Mexico
		Burma, Malaysia, Thailand	Tiger beetle	Pupa	China, Japan, Thailand
		Australia, Africa, South	Locust lang	Egg, adult	Australia, Africa
		America	Psylla head lice	Larva	
Ants	Adult	Indonesia, India, Thailand, Australia, Mexico, America	Head lice fleas	Adult	Indonesia, Vietnam, Africa, Mexico, South America
Hydrophilid	Adult	China, Japan			
Bagworm	Overwintering larvae	China	Red current worm	Overwintering larvae	China

Table 1. Edible insects from all over the world.

2. Development and utilization of edible insects resource

Edible insects can be divided into food insect, drug/medicinal insects and drug dual-use insects, etc., based on the different insects eaten resources classification. Edible insects are directly for a daily food consumption and the insect has important nutritional value to human to be developed and utilized. The U.N.'s Food and Agriculture Organization (FAO) has released a report in 2013 called Edible Insects: Future Prospects for Food and Feed Security [6]. It outlines the many benefits of eating insects—for human in the entire world. As early as 1980, it was put forward to supplement the human food shortage in the Fifth Latin American congress of Dietitians and Nutritionists, which should regard them that as part of food source insects as human food in many countries has been increasingly apparent at present.

Scientists have found that some insect protein from the red ants, grasshoppers and some of predaceous diving beetle (Dytiscidae)are enough to compete with lean beef.

The protein of adult insect content is rich, significantly higher than that of pork, beef, chicken, fish and eggs. Experts predict that insects will be the third category only after cell raw material and microbial protein sources in future [7]. As in poor regions, people need essential nutrients to provide, the services of the services of insect and spider equally good. And as in developed country like the USA, the insect and spider are the higher protein food from a healthy choice. Insects are a highly nutritious and healthy food source with high fat, protein, vitamin, fiber and mineral content [6]. "Gathering and farming insects can offer employment and cash income either at the household level or in larger industrial-scale operations." It could offer work to millions of people around the world. Besides, there is evidence that most of breeding insects produced the harmful greenhouse gases to the environment that may be less than those of livestock [8]. This result will help to decrease the cost of food production, reduce emissions from greenhouse gas. In recent years, along with the progress of modern science and technology, the process technology of the functional food and health-food markets of edible insects accelerates unprecedentedly in China. For example, concentrated insect protein oral liquid specialized in honey, royal jelly, pollen and propolis, the traditional shellac ash, etc. Insect oils of some are mainly used as the functional fat-soluble ingredients.

As mentioned above, insects are the largest species and biological organisms on earth. And their resources arre extremely rich. Moreover, its nutrition structure is reasonable with high nutritional value and many functional constituent; insects thus will be considered as an important food source for the development and utilization with huge potentials. Combined with the existing research results and their advantages of high reproductive growth speed and low feed cost, large-scale production provides a reliable technical support. The development and utilization of edible insect resources thus has a very broad prospect.

3. The value of nutrition and health benefits of edible insects

According to the records, about 3650 species of the edible insects have been found and used [8]. The advantages of edible insects in the food development, one highlights show, at the beginning of competing namely, already competed from widely distributed, best variety, fast reproduction, high nutritional value, breeding easy and high food conversion rate indistinct development competes for the sustainability in ecosystems and biodiversity. They thus have become an additional source of food with high nutritional value containing rich protein. At the same time, the insects' food has low fat and low cholesterol with a reasonable structure (less fleshy fiber) easily absorbed, and abundant trace elements, etc. It is, therefore, better than that of meat and eggs [9].

The results of research showed that edible insect contains various nutrient elements, such as protein, amino acids, fat, fatty acid, vitamins and mineral elements.

3.1. Edible insect protein

Onincx said "It proves the hypothesis that insects can be a more efficient source of protein, and I definitely believe there is a future for edible insects" [10]. Studies have shown that protein content is not the same for the different insect states. The protein content of adults is the highest, pupa is at second and larvae at the lowest. For wasps, the adult has 71.07%, pupa has 58.59% and a larva has 50.83% of protein content, according to the protein calculations of insects at different ages.

The protein content of different subjects' insect is not the same either. The Orthoptera is higher than Homoptera, higher than Odonata, Diptera, Hymenoptera, Hemipter, Lepidoptera and then Coleoptera insect [11].

Amino acid is the basic functional unit for biological macromolecular protein, and is also an important part of food constituted as insects' nutrition. The amino acid content of edible insects is 10–70%, and essential amino acid content is 10–30%. Most of the amino acid ratios for insect are appropriate and have approached or even exceeded that of the WHO/FAO required ratio. Study also revealed that the existence of large amount of free amino acids associated with insect freshness [6, 12]. The content of free amino acid of edible insects in the blood is about 3000–23400 mg/kg that is higher than any other higher animals of the universe.

3.2. Carbohydrate of edible insects

In addition to glucose, triose, glycogen, erythritol, ketose sugar, fructose and ketoheptose, edible insects' carbohydrates (sugars) sort are very rich and the sea algae sugar (insects' ingredient blood sugar) content is the same. Edible insects are easy to digest and absorb carbohydrates, and total sugar containing amount is generally as low as 1–10% or even lower [13]. For example, the total sugar content of *Cyclopelta parva* is 1.45%, and that of *Tessaratoma papillosa* is 0.15% [14].

Chitin is also the main material of edible insects' skin and bones. Its chemical name is N-acetyl-D-glucosamine copolymer with the function of adsorption abilities for a specific toxin. then, It is also a low calorie food that it has the very high nutrition value for the good for the health care. Chitin promotes intestinal peristalsis, fine regulating intestinal bacteria, reduce weight due to fat, anti-aging, enhance the immune function and assist in preventive treatment of high blood pressure, etc. Chitin is usually between 15 and 18%, rich in edible insect body. However, at different insect states, chitin content is different, such as the chitin content of dry silkworm pupa is 3.73% and Skim pupa's content reached 5.55% [3].

3.3. Mineral elements and vitamin of edible insects

Edible insects are rich in mineral elements, including Ca, P, Fe and zinc, etc., which are often needed as the supplement of human body. It is reported that feed insects can fulfill the requirement of animals' Fe, Cu, Zn and Mg mineral elements [15]. The locusts contain 27 kinds of mineral elements, notably Mn, Fe, Cu and Zn [16]. Many ants are rich in Zn, Se, Mn and Mg, etc. The level of Zn is two times more than pork liver, and eight times higher than soybean [17]. In addition to the constant element, edible insects are rich in Se, Co, Ni and Cd trace elements. The Chinese rice locust and yellow powder insect have Se content at 4.62 and 4.75 mg/kg,respectively. The Se element can accelerate detoxification, inhibit carcinogenic activity, destroy the carcinogen and prevent cancer cell growth and division. Other elements' content, such as Ni 1.22 mg/g, Co 1.36 mg/g and Cr 1.52 mg/g are reported in *Formica (Coptoformica) mesasiatica* Dlussky [3].

There are numerous vitamins in insect body, mainly including vitamin B1 (thiamine), B2 (riboflavin), B3 (niacin), B6 (pyridoxine), C, D, E, K and carotene. *Macrotermes annandalei* contains vitamin A 25.0, vitamin D 85.4 and vitamin E 11.7 I.U./g (International Unit/g). Vitamins are essential substances for maintaining normal physiological function of human body.

3.4. Lipid substances of edible insects

Insects are rich in oil and lipids. The fat content of pupae and larvae is higher than the adult insect. Report shows a decline in fat content after feather state of the insect. Fat content is commonly between 10 and 50% for edible insects. A recent study has found the fat content of wasp. In larvae, the fat content is 29.01%, while in pupae, it is 27.25% and in adults, it is for 17.22% [10]. Unsaturated fatty acid and palmitic acid are higher in edible insects. Among them, linolenic acid content is higher in Lepidoptera, and the oil acid content is higher in Coleoptera.

Related research also proved the variation in fat content for different species of insects. It is revealed that the fat content of Coleoptera is higher than that of Lepidoptera, Half wings, Hymenoptera, Odonata, Diptera and Orthoptera [19]. In fact, natural insect wax also contains a small amount of senior fatty acids. They can be used as pharmaceutical raw materials such as textile and stencil making model. Based on the infrared spectra analysis of insect wax, it was suggested that insect's wax is composed of mainly long-chain hydrocarbons, fatty alcohols, fatty acids and some compounds with aromatic rings mixture.

4. The nutritional evaluation of insect oil/fat

Insect oils (fat) are a kind of nutrient substances with several physiological and biological activities and functions. It has a high value of research, development and utilization no matter whatever be the quantity or quality. Fat content of insects' body changes with its life cycle, meanwhile it is closely related to the growing up of the insect species [20].

4.1. The oils and fats content in insects' body

Many studies have also displayed that the fat content of insects differed in the same species. The pupa and larva's oils (fat) were higher than adults' in the same species. And, during the winter period, the insects' oil contents were higher. The fat content of insects' dry body was commonly 10%, while many other insects have fat content of 30%, or even up to 77.16% (**Table 2**).

4.2. Fatty acid composition of insect oils and fats

Insects are rich in fat and their fatty acid composition is reasonable. The saturated fatty acids and unsaturated fatty acid ratio of edible insect is generally less than 0.4. Its partial fatty acid

Insect species	Crude fat %	Insect species	Crude fat %	Insect species	Crude fat %
Locates moth larvae	77.17	*Desert locust*	17	*Bollworm* larvae	49.48
		Asian corn borer larvae	46.08	The big spot *well-known*	14.5
Clanis bilineata	15.44	*Verdigris beetle* larvae	14.05	*Aprioma germari Hope*	41.46
Moths' larva	32.26			*stratiomyiid*	13.93
Mylabris cichorii	13.96	*Yellow mealworm* adult	19.23	*Pink neck sawyer* larve	35.89
Tussah male adult	39.49	*Oxya chinensis*	8.24	*Musca domestica* pupa	10.55
Cabbage worm	11.8	*Polyrhachis vicina* female	9.5	*Tussah* pupa	31.25
Tenebrio molitor L	28.8–34.0	*Holotrichia oblita* Fald	29.84	*Redchest* prickly ant adult	8.53
Polyrhachis vicina adult	8.57	*Housefly* larvae	12.61	*Convex star* flowers beetle larvae	19.35
Macrotermes annandalei	28.3	*Star longhorn* beetle larvae	35.19	*Green hsu* well-known	7.5
Epicauta chinensis	8.22	*Atractomorpha sinensis*	2.87–4.91	*Tussah cicada* larva	2.63
Wood stupid worm	26.46				
Acrida cinerea	2.89				

Table 2. Crude fat (dry weight) content (%) of some insects.

composition ratio is close to the fatty acid composition of fish proportion, and thus can be used as a natural health care product. The saturated fatty acids (SFA) of insects is mostly composed of the palmitic acid (C16:0) but not stearic acid (C18:0) which is relatively high in vertebrates. In addition, insect oil has the odd number carbon fatty acids rarely existing as the pentadecanoic acid and heptadecanoic acid that are relatively rare in the nature but extremely common in insects. As shown in **Table 2**, the heptadecanoic acid content of termites' adult, the housefly larvae and housefly adults were all above 2%. As odd number carbon fatty acid has special raw active function, it was found that they have stronger antitumor activity. Therefore, many researchers are very interested in insects' enrichment and separation of odd number carbon fatty acids, leading a hotspot in the research of the insect oil.

The insect oil is a solvent of natural active products containing the lecithin and fat-soluble D raw element (such as vitamin A, D, E). These active natural products have a strong physiological and biological function with the extremely important value.

	Composition of fatty acid										
Insect	**14:0**	**15:0**	**16:0**	**17:0**	**18:0**	**16:1**	**17:1**	**$18:1_{n-9}$**	**$18:2_{n-6}$**	**$18:3_{n-3}$**	**>18**
Larvae of *Tenebrio molitor*	0.51	0.99	23.6	–	1.4	1.8	2.7	44.7	24.1	1.5	–
House fly larvae	2.2	–	19.7	3.2	2.3	12.7	1.0	18.2	32.5	3.3	0.2
House fly adult	3.5	0.5	15.6	3.4	4.8	5.7	–	26.8	35.4	–	4.5
House fly pupa	0.7	2.1	27.6	–	2.2	5.8	14.8	18.3	14.9	2.1	0.2
Silkworm pupa	–	–	30.0	–	7.5	–	–	25.6	10.9	26.0	–
The termites adult	0.6	1.0	31.0	2.6	3.4	1.0	0.6	9.5	43.1	3.0	4.2

Table 3. The fatty acid composition of some of insects' oil and fats (%).

4.3. The nutrition evaluation of insect oils

The fatty acid of insect oil and fats predominantly composes of unsaturated fatty acid and its proportion could be more than 60%, and can even reach up to 80% in some insects. The fatty acid of insect oil and fats predominantly composes of unsaturated fatty acid. And, its proportion could be more than 60%, and can even reach up to 80% in some insects. The fatty acid composition of some of the insects' oil and fats is shown in **Table 3**.

Among the unsaturated fatty acids (USFA) and monounsaturated fatty acids (MUFA) of insects, many oleic acid (C18:1) is a unique composition with the proportion at about 30 ± 10% or more. This is much close to fish food and better than poultry meat and eggs [5, 18, 20]. It is good for the health of human body. The bioactive mechanism of insects' fatty acid has been studied in the body and has proven the exact functional fatty acids physiological activity. It pointed out that the high linoleic acid content in certain insects has a close relationship with its strong reproductive functions. Research indicated that stearic acid could neutralize the nutritional effects, and the lauric acid (C12:0) and nutmeg acid (C14:0) can lead to the elevated level of the fatty acid of cholesterol. These kinds of saturated fatty acids (SFA) are low in insect oil.

Nutritionists believe that general proportion between the n-3 and n-6 (PUFA) should be based on the breast for 1:3–10 [21]. According to the best food and nutritional science, linoleic acid and linolenic acid are the essential fatty acids (EFA). These two kinds of fatty acids are obtained only from food directly. In case EFA is lacking, the oil should be restricted to oil's fatty acid with the low biological titer, and nutritional value is thus low. It can directly cause the growth retardation, reproductive barriers, skin damage (such as a skin rash), liver, kidney, nerve and visual diseases. The over-taken polyunsaturated fatty acids (PUSFA) can cause chronic hazards. Therefore, the world health organization (WHO) recommended a standard of food oil in adults, the recommended dietary fatty acids taken (1990) are produced by the percentage of the total energy, energy and food: fat 15–30%, of them SFA < 10, PUFA 3–7 [21].

In a word, insects have the characters of quick propagation, high content of fat composed of reasonable composition of fatty acids. They are thus a good edible oil resource with high quality.

5. The secondary metabolites of edible insects and potential medicinal substances

It has been proved by a large number of studies in recent years that the insect secondary metabolite is important sources to find new leading compounds. Arthropod natural products with insect constituent are structurally diverse, including compounds derived from fatty acid, polyketide, terpenoid, nucleoside and amino acid pathways. However, the biosynthesis of most of these compounds has not been studied in detail (**Figure 1**) [22]. Historically,

http://dx.doi.org/10.5772/intechopen.70085

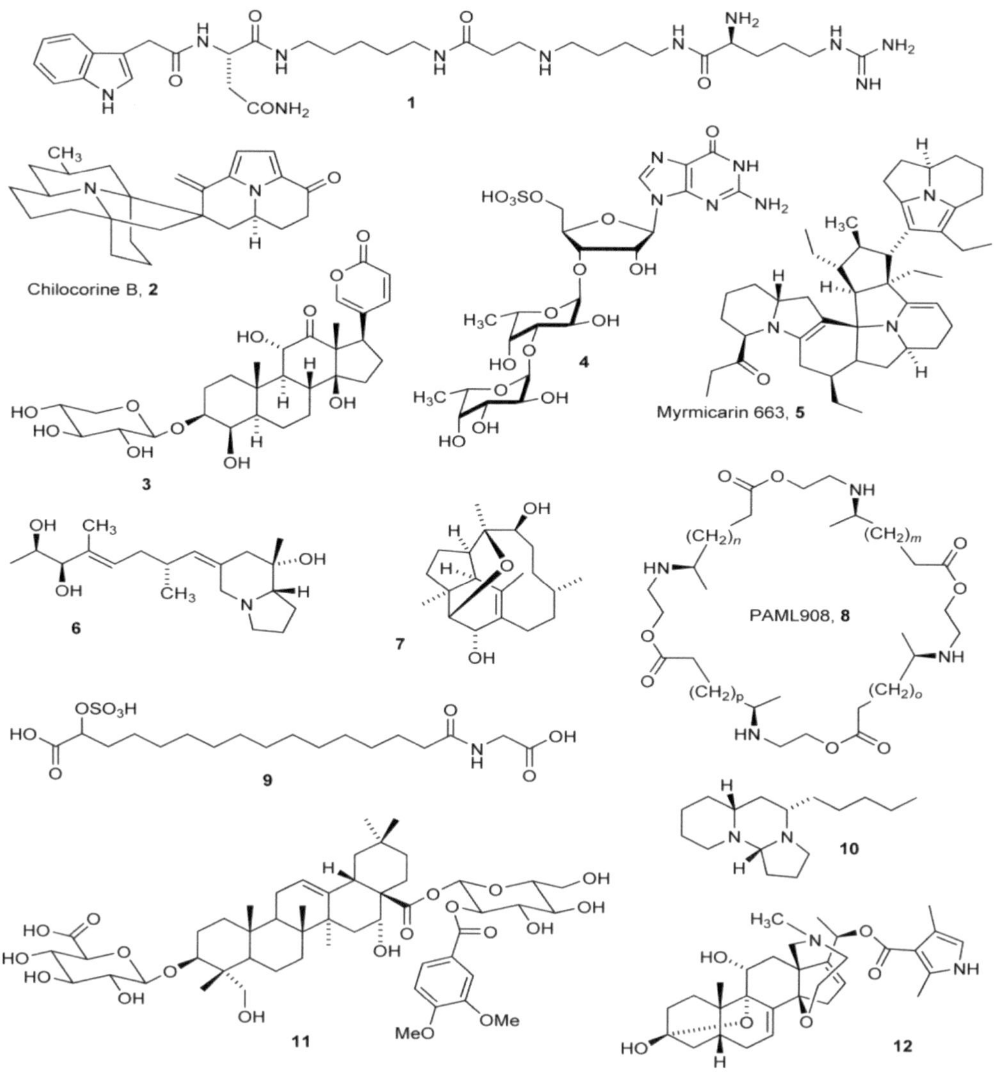

Figure 1. Examples of arthropod natural products from spiders (1, 4), mites (6), ants (5, 7, 10), fireflies (3), termites (7), grasshoppers (9), and beetles (2, 8, 11, 12).

traditional use of plants as medicines, known as "ethnobotany", has been extensively recognized and studied. It is worth noting that insects have been utilized as medicines in diverse cultures, especially in traditional Chinese medicines. It may be valuable for the development of the useful drugs. Another ongoing investigation by our group was the searching of new antibacterial structure from insects' natural products. More modern studies designed to determine the medicinal properties of isolated chemical components from insects and other arthropods will be performed.

6. Edible insects in China

According to the results of the survey, edible insect resources have been reported for more than 283 species with subspecies, involving 13 orders, 71 families in China [23]. Among them, the Orthoptera, Hemiptera, Coleoptera, Lepidoptera and Hymenoptera account for the majority number. Though most of the species have not been reported, the quantity of edible species identified is still growing. For this purpose, the author studied more than 283 species and subspecies of the edible insects in China. We recorded them as follows with * marked for the first-time report. China's edible insect species and their edible insects were discussed as follows:

1. Ephemerida

The order includes two families of the larva of Ephemerellidae and *Ephemerella jinghongensis* Xu et al.

2. Odonata

The order includes four families of Aeshnidae, Gomphidae, Libellulidae and Lestidae.

2.1. *Aeschnidae

The larva of **Anax parthenope julius* Brauer.

2.2. Gomphidae

The larva of *Gomphus cuneatus* Needham.

2.3. Libelluidae

1. The larva of *Crocothemis servilia* Drury.
2. The larva of * *Orthetrum albistylum* Selys.
3. The larva of * *Orthetrum triangula remelania* Selys.
4. The larva of * *Pantala flavescens* Fabricius.
5. The larva of * *Sympetrum uniforme* Selys.

2.4. Lestidae

The larva of *Lestes praemorsa* Sel.

3. Blattodea

The order includes two families of Blattidae and Corydiidae.

3.1. Blattidae

1. The nymphal/adult of *Periplaneta americana* L.
2. The nymphal/adult of *Periplaneta australasiae* L.

3.2. Corydiidae

The nymphal/adult of *Eupolyphaga sinensis* Walker.

4. **Mantodea**

The order has only the Mantidae familie.

1. The nymphal/adult of *Mantis religiosa* L.
2. The nymphal/adult of *Paratenodera sinensis* Saussure.
3. The nymphal/adult of *Statilia maculata* Thunberg.
4. The nymphal/adult of *Tenodero bravico* Beier.
5. The nymphal/adult of *Tenodero sinensis* Saussure.

5. **Isoptera**

The order includes two families of Rhinotermitidae and Termitidae.

5.1. Rhinotermitidae

The larva/nest/adult of *Coptotermes formosanus* Shiraki.

5.2. Termitidae

1. The larva/nest/adult of *Macrotermes acrocephalus* Ping.
2. The larva/nest/adult of *Macrotermes annandalei* Slivestri.
3. The larva/nest/adult of *Macrotermes barneyi* Light.
4. The larva/nest/adult of *Macrotermes denticulatus* LietPing.
5. The larva/nest/adult of *Macrotermes jinhongensis* PingetLi.
6. The larva/nest/adult of *Macrotermes menglongensis* Han.
7. The larva/nest/adult of *Macrotermes yunnanensis* Han.
8. The larva/nest/adult of *Odontotermes angusti gnathus* TsaietChen.
9. The larva/nest/adult of *Odontotermes annulicornis* XiaetFan.
10. The larva/nest/adult of *Odontotermes conignathus* XiaetFan.
11. The larva/nest/adult of *Odontotermes formosanus* Shiraki.
12. The larva/nest/adult of *Odontotermes foveafrons* XiaetFan.
13. The larva/nest/adult of *Odontotermes gravelyi* Silvestri.
14. The larva/nest/adult of *Odontotermes hainanensis* Light.
15. The larva/nest/adult of *Odontotermes yunnanensis* TsaietChen.

6. **Orthoptera**

The order includes four families of Acridiidae, Gryllidae, Gryllotalpidae and Tettigoniidae.

6.1. Acridiidae

1. The nymphal/adult of *Acrida chinensis* Westwood.
2. The nymphal/adult of *Acrida. oxycephala* Pallas.
3. The nymphal/adult of *Acrida. turrita* L.
4. The nymphal/adult of *Arcyptera fusca* Pall.
5. The nymphal/adult of *Atractomorpha sinensis* Boliver.
6. The nymphal/adult of *Bryodema gebleri* FisherWaldheim
7. The nymphal/adult of *Calliptamus abbreviatus* Ikonn.
8. The nymphal/adult of *Calliptamus italicus* L.
9. The nymphal/adult of *Calliptamus barbaruscephalates* FisherWaldheim
10. The nymphal/adult of *Ceracris kiangsu* Tsai.
11. The nymphal/adult of *Chondracris rosea* DeGeer.
12. The nymphal/adult of *Dociostaurus kraussini grogeniculatus* Tar.
13. The nymphal/adult of *Gomphocerus sibiricus* L.
14. The nymphal/adult of *Locusta migratoria manilensis* Meyen.
15. The nymphal/adult of *Locusta migratoria migratoria* L.
16. The nymphal/adult of *Oedaleus decorus* Germ.
17. The nymphal/adult of *Oxya chinensis* Thunberg.
18. The nymphal/adult of *Oxya intericata* Stal.
19. The nymphal/adult of *Oxya japonica* Thunberg.
20. The nymphal/adult of *Pararcyptera microptera* FisherWaldheim.
21. The nymphal/adult of *Patanga japonica* Bolivar.
22. The nymphal/adult of *Skirakiacris shirakii* Bolivar.
23. The nymphal/adult of *Sphingonotus* spp.
24. The nymphal/adult of *Stauroderus scalaris* FisherWaldheim.

6.2. Gryllidae

1. The adult of *Brachytrupes portentosus* L.
2. The adult of *Gryllulus bimaculatus* DeGeer
3. The adult of *Gryllulus chinensis* Weber
4. The adult of *Gryllulus testaceus* Walker

5. The nymphal/adult of *Teleoqzyllus derelictus* Gorochov.
6. The nymphal/adult of *Tarbinskiellus portentosus* (Lichtenstern).

6.3. Gryllotalpidae

1. The adult of *Gryllotalpa africana* Palisotde Beauvojs.
2. The adult of *Gryllotalpa orientalis* Burmeister.
3. The adult of *Gryllotalpa unispina* Saussure.

6.4. Tettigoniidae

The nymphal/adult of *Damalacantha vacca sinica* B. Bienko.

7. Homoptera

The order includes five families of Cicadidae, Coccidae, Flatidae, Membracidae and Pseudococcidae.

7.1. Cicadidae

1. The nymphal of *Cicada flammata* Dist.
2. The nymphal of *Cryptotympana atrata* Fabr.
3. The nymphal of *Platypleura kaempferi* Fabr.

7.2. Coccidae

The egg/adult of *Ericerus pela* Chavanness.

7.3. Flatidae

The nymphal of *Lawana imitata* Melichar.

7.4. Membracidae

The nymphal/adult of *Darthula hardwicki* Gray.

7.5. Pseudococcidae

The nymphal of *Phenacoccus prunicola* Borchs.

8. Hemiptera

The order includes five families of Belostomatidae, Coreidae, Corixidae, Noronectidae and Pentatomidae.

8.1. Belostomatidae

1. The nymphal/adult of *Kirkaldgia degrollei* Vuillefro.
2. The nymphal/adult of *Lethocerus indicus* Lepeletieret Serville.
3. The nymphal/adult of *Sphaerodema rustica* Fabricius.

8.2. Coreidae

The nymphal/adult of *Mictis tenebrosa* Fabricius.

8.3. Corixidae

1. The nymphal/adult of *Micromecta quadriseta* Lundblad.
2. The nymphal/adult of *Sigara substriata* Uhler.

8.4. Noronectidae

1. The nymphal/adult of *Anisops fieberi* Kirkaldy.
2. The nymphal/adult of *Enithares sinica* Stal.
3. The nymphal/adult of *Notonecta chinensis* Fallou.

8.5. Pentatomidae

1. The nymphal/adult of *Coridicus chinensis* Dallas.
2. The nymphal/adult of *Cyclopelta parva* Distant.
3. The nymphal/adult of *Erthesina fullo* Thunberg.
4. The nymphal/adult of *Eurostus validus* Dallas.
5. The nymphal/adult of *Eusthenes curpreus* Westwood.
6. The nymphal/adult of *Eusthenes saevus* Stal.
7. The nymphal/adult of *Nezara viridula* L.
8. The nymphal/adult of *Tessara toma papillosa* Drury.

9. Coleoptera

The order includes 15 families of Anobiidae, Bruchidae, Buprestidae, Cerambycidae, Crioceridae, Curculionidae, Dynastidae, Dytiscidae, Getoniidae, Hydrophilidae, Melolonthidae, Rutelidae, Scarabaeidae, Scolytidae and Tenebrionidae.

9.1. Anobiidae

The nymphal of *Lasioedrma serricorne* Fabricius.

9.2. Bruchidae

1. The nymphal/adult of *Bruchus pisorum* L.
2. The nymphal/adult of *Bruchus rufimanus* Boheman.

9.3. Buprestidae

1. The nymphal of *Chalcophora yunnana* Fairmaire.
2. The nymphal of *Coraebus sidae* Kerremans.

3. The nymphal of *Coraebus sauteri* Oben.
4. The nymphal of *Sphenoptera kozlovi* B. Jak.

9.4. Cerambycidae

1. The nymphal/adult of *Anoplophora chinensis* Forster.
2. The nymphal/adult of *Anoplophora nobilis* Ganglbauer.
3. The nymphal/adult of *Apriona germari* Hope.
4. The nymphal/adult of *Aromia bungii* Faldermann.
5. The nymphal/adult of *Stromatium longicorne* Newman.
6. The nymphal/adult of *Psacothea hilaris* Pascoe.

9.5. Crioceridae

The nymphal/adult of *Sagra femorata purpurea* Lichtenstein.

9.6. Curculionidae

1. The nymphal/adult of *Cyrtotrachelus bugueti* Guer.
2. The nymphal/adult of *Cyrtotrachelus longimanus* Fabricius.
3. The nymphal/adult of *Macrochirus longipes* Drury.
4. The nymphal/adult of *Otidognathus davidis* Fabricius.

9.7. Dynastidae

1. The nymphal/adult of *Allomyrina dichotoma* L.
2. The nymphal/adult of *Oryctes rhinoceros* L.

9.8. Dytiscidae

1. The adult of *Cybister japonicus* Sharp.
2. The adult of *Cybister. limbatus* Fabricius.
3. The adult of *Cybister ripunctatus* Olivier.
4. The adult of *Eretes stictius* L.

9.9. Getoniidae

1. The nymphal/adult of *Dicranocephalus wallichi bowringi* Pascoe.
2. The nymphal/adult of *Oxycetonia jucunda* Faldermann.
3. The nymphal/adult of *Protaetia aerata* Erichson.

9. 10. Hydrophilidae

1. The adult of *Hydrophilus acuminatus* Motsch.
2. The adult of *Hydrous acuminatus* Motsch.
3. The adult of *Hydrous hastatus* Herbst.

9. 11. Melolonthidae

1. The nymphal/adult of *Holotrichia diomphalia* Bates.
2. The nymphal/adult of *Holotrichia lata* Brenske
3. The nymphal/adult of *Holotrichia oblita* Faldermann.
4. The nymphal/adult of *Holotrichia ovata* Chang.
5. The nymphal/adult of *Holotrichia parallela* Motsch.
6. The nymphal/adult of *Holotrichia sinensis* Hope.
7. The nymphal/adult of *Holotrichia srobiculata* Brenske.
8. The nymphal/adult of *Holotrichia szechuanensis* Chang.
9. The nymphal/adult of *Polyphylla laticollis* Lewis.

9. 12. Rutelidae

The nymphal/adult of *Anomala corpulenta* Mots.

9. 13. Scarabaeidae

The nymphal/adult of *Catharsiusmolossus* L.

9. 14. Scolytidae

1. The nymphal of *Sphaerotrypes yunnanensis* Tsaiet Yin.
2. The nymphal of *Tomicus piniperd* L.
3. The nymphal of *Xyleborus emarginatus* Eichhoff

9. 15. Tenebrionidae

1. The nymphal/pupa of *Tenebriomolitor* L.
2. The nymphal/pupa of *Tenebriomolitor obscurus* Feb.
3. The nymphal/pupa of *Tribolium confusum* Jac.du Val.

10. Megaloptera

The order has only the Corydalidae family.

The nymphal of *Acanthacoryda lisorientalis* Mclachlan.

11. Lepidoptera

The order includes twenty-one families of Hesperiidae, Papilionidae et al.

11. 1. Hesperiidae

1. The pupa of *Erionota torus* Evans.
2. The pupa of *Parnara guttata* Bremeret Gray.

11. 2. Papilionidae

1. The pupa of *Papilio machaon* L.
2. The pupa of *Papilio polytes* L.
3. The pupa of *Papilio xuthus* L.

11. 3. Pieridae

The pupa of *Pieris rapae* L.

11. 4. Satyridae

The pupa of *Mycalesis gotoma* Moore.

11. 5. Aegeriidae

1. The larva/pupa of *Paranth reneregalis* Butler.
2. The larva/pupa of *Parathene tabaniformis* L.

11. 6. Bombycidae

1. The pupa of *Andraca bipunctata* Walker.
2. Silkworm chrysalis and silk moth of *Bombyx mori* L.
3. The pupa of *Theophila mandarina* Moore.

11. 7. Carposinidae

The pupa of *Carposina niponensis* Walsingham.

11. 8. Cossidae

1. The larva of *Cossus chinesis* Rothschild
2. The larva of *Cossus cossus* L.
3. The larva of *Cossus hunanensis* Daniel.

11. 9. Eucleidae

1. The pupa of *Cania bilineata* Walke.
2. The pupa of *Thosea sinensis* Walker.

11. 10. Gelechiidae

1. The larva/pupa of *Pectionophora gossyeilla* Saunders.
2. The larva/pupa of *Platyedra gossypiella* Saunders.

11. 11. Geometridae

The larva/pupa of *Biston marginata* Matsmura.

11. 12. Hepialidae

1. The larva/pupa of *Hepialus albipictus* Yang.
2. The larva/pupa of *Hepialus altaicola* Wang.
3. The larva/pupa of *Hepialus armoricanus* Oberthur.
4. The larva/pupa of *Hepialus baimaensis* Liang.
5. The larva/pupa of *Hepialus cingulatus* Yang et Zhang.
6. The larva/pupa of *Hepialus deudi* Poujade.
7. The larva/pupa of *Hepialus deqinensis* Liang.
8. The larva/pupa of *Hepialus dongyuensis* Liang.
9. The larva/pupa of *Hepialus ferrugineus* Li, Yang et Shen.
10. The larva/pupa of *Hepialus ganna* Hubner.
11. The larva/pupa of *Hepialus gonggaensis* FuetHuang.
12. The larva/pupa of *Hepialus jinshaensis* Yang.
13. The larva/pupa of *Hepialus kangdingensis* Chu et Wang.
14. The larva/pupa of *Hepialus kangdingroides* Chu et Wang.
15. The larva/pupa of *Hepialus lijiangensis* Chu et Wang.
16. The larva/pupa of *Hepialus litangensis* Liang.
17. The larva/pupa of *Hepialus luquensis* Yang et Yang.
18. The larva/pupa of *Hepialus macilentus* Lversmann.
19. The larva/pupa of *Hepialus markamensis* Yang, Li et Shen.
20. The larva/pupa of *Hepialus meiliensis* Liang.
21. The larva/pupa of *Hepialus menyuanensis* Chu et Wang.
22. The larva/pupa of *Hepialus nebulosus* Alpheraky.
23. The larva/pupa of *Hepialus oblifurcus* Chu et Wang.
24. The larva/pupa of *Hepialus pratensis* Yang.

25. The larva/pupa of *Hepialus renzhiensis* Yang.
26. The larva/pupa of *Hepialus H.sichuanus* Chu et Wang.
27. The larva/pupa of *Hepialus varians* Staudinger.
28. The larva/pupa of *Hepialus xunhuaensis*Yang et Yang.
29. The larva/pupa of *Hepialus yeriensis* Liang.
30. The larva/pupa of *Hepialus yulongensis* Liang.
31. The larva/pupa of *Hepialus yunlongensis* Chu et Wang.
32. The larva/pupa of *Hepialus yunnanensis* Yang et Li.
33. The larva/pupa of *Hepialus yushuensis* Chu et Wang.
34. The larva/pupa of *Hepialus zhangmoensis* Chu et Wang.
35. The larva/pupa of *Hepialus zhayuensis* Chu et Wang.
36. The larva/pupa of *Hepialus.zhongzhiensis* Liang.
37. The larva/pupa of *Napialus hunanensis* Chu et Wang.

11. 13. Lasiocampidae

1. The adult/pupa of *Dendrolimus houi* Lajonquiere.
2. The adult/pupa of *Dendrolimus kikuchii* Matsumura.
3. The adult/pupa of *Dendrolimus punctatus* Walker.
4. The adult/pupa of *Dendrolimus punctatus wenshanensis* Tsai et Liu.
5. The adult/pupa of *Dendrolimus superans* Butler.

11. 14. Noctuidae

1. The pupa of *Agrotis ipsilon* Pottemberg.
2. The pupa of *Anomis flava* Fabr.
3. The pupa of *Heliothis armigera* Hubner.
4. The pupa of *Hydrillodes morosa* Butler.
5. The pupa of *Laphygma exigua* Hubner.
6. The pupa of *Leucania separata* Walker.
7. The pupa of *Naranga aenescens* Moore.
8. The pupa of *Prodenia litura* Fabr.
9. The larva/pupa of *Sesamia inferens* Walker.

11. 15. Notodontidae

1. The adult/pupa of *Leucodonta bicoloria* Denis et Schiffermuller.
2. The adult/pupa of *Notodonta dembowskii* Oberthuer.
3. The adult/pupa of *Phalera assimilis* Bremer et Gray.
4. The adult/pupa of *Phalera bucephala* L.
5. The adult/pupa of *Semidonta biloba* Oberthuer.

11. 16. Psychidae

The larve/pupa of *Psychidae* spp.

11. 17. Pyralidae

1. Insect tea of larva feces for *Aglossa dimidiata* Haworth.
2. The larva/pupa of *Chilo suppressalis* Walker.
3. The larva of *Chilo fuscidentalis* Hampson.
4. The larva/pupa of *Chilo* sp. (English named, Bamboo maggots)
5. The pupa of *Cnaphalocrocism edinalis* Guenée.
6. The pupa of *Dichocrocis punctiferalis* Guenée.
7. The larva/pupa of *Ostrinia furnalis* Guenée.
8. The pupa of *Plodia interpunctella* Hubner.
9. The pupa of *Sylepta derogata* Fabr.
10. The larva/pupa of *Tryporyza incertulas* Walker.

11. 18. Saturniidae

1. The larva/pupa of *Antheraea pernyi* Geurin.
2. The larva/pupa of *Philosamia cynthia* Drury.

11. 19. Sphingidae

1. The larva/pupa of *Clanis bilineata* Walker.
2. The larva/pupa of *Clanis deucalion* Walker.
3. The larva/pupa of *Herse convolvuli* L.
4. The larva/pupa of *Smerithus plannus* Walker.

11. 20. Tortrcidae

The pupa of *Leguminivora glycinivorella* Matsumura.

11. 21. Xyloryctidae

1. The pupa of *Linoclostis gonatias* Meyrick.
2. The pupa of *Xyloryctidae* spp.

12. Diptera

The order includes three familie of Muscidae, Sarcophagidge and Tipulidae.

12. 1. Muscidae

The egg/larva of *Musca domestica* L.

12. 2. Sarcophagidge

The larva of *Sarcophagidae* spp.

12. 3. Tipulidae

The larva of *Tipula paltudosa* Meig.

13. Hymenoptera

The order includes seven families of Agaonidae, Apidae, Polistidae, Sco1iidae, Sphecidae, Vespidae and Formicidae.

13. 1. Agaonidae

The egg/larva/pupa/adult of *Blastophaga pumilae* Hill.

13. 2. Apidae

1. The larva/pupa of *Apis cerana* Fabricius.
2. The larva/pupa of *Apis mellifera* L.
3. The larva/pupa of *Bombus speciosus* Smith.
4. The larva/pupa of *Megapis dorsata* Fabricius.
5. The larva/pupa of *Megapis florae* Fabricius.

13. 3. Polistidae

1. The larva/pupa of *Polistes antenalis* Perez.
2. The larva/pupa of *Polistes chinensis* Fabricius.
3. The larva/pupa of *Polistes gigas* Kirby.
4. The larva/pupa of *Polistes hebraeus* Fabricius.
5. The larva/pupa of *Polistes mandarimnus* Saussure.
6. The larva/pupa of *Polistes.sagittarius* Sassuer.
7. The larva/pupa of *Polistes salcatus* Smith.

13. 4. Sco1iidae

The larva/pupat of *Scoliidae* spp.

13. 5. Sphecidae

The larva/pupat of *Sphecidae* spp.

13. 6. Vespidae

1. The larva/pupa of *Provespa barthelemyi* Buysson.
2. The larva/pupa of *Vespa analis* Buysson.
3. The larva/pupa of *Vespa basalis* Smith.
4. The larva/pupa of *Vespa bicolor bicolor* Fabricius.
5. The larva/pupa of *Vespa crabro* L.
6. The larva/pupa of *Vespa ducalis* Smith.
7. The larva/pupa of *Vespa mandarinia* Smith.
8. The larva/pupa of *Vespa sorror* Buysson.
9. The larva/pupa of *Vespa tropica ducalis* Smith.
10. The larva/pupa of *Vespa variabilis* Buysson.
11. The larva/pupa of *Vespa velutina auraria* Smith.
12. The larva/pupa of *Vespa* spp.

13. 7. Formicidae

1. The egg/larva/pupa/adult of *Camponotus japonicus* Mayer.
2. The egg/larva/pupa/adult of *Carebara lignata* Westwood.
3. The larva/pupa of *Formica rufa* L.
4. The larva/pupa of *Formica aquilonia* Yarrow.
5. The larva/pupa of *Formica beijingensis* Wu.
6. The larva/pupa of *Formica fusca* L.
7. The larva/pupa of *Formica japonica* Mottschulsky.
8. The larva/pupa of *Formica sanguinea* Latr.
9. The larva/pupa of *Formica uralensis* Ruzsky.
10. The larva/pupa of *Formica yessensis* Forel.
11. The larva/pupa of *Lasius flavus* Fabricius.
12. The larva/pupa of *Oecophylla smaragdina* Fabricius.
13. The egg/larva/pupa/adult of *Polyrhachis dives* Smith.

14. The larva/pupa of *Polyrhachis illaudata* Walker.

15. The egg/larva/pupa/adult of *Polyrhachis lamellidens* Smith.

16. The egg/larva/pupa/adult of *Polyrhachis vicina* Roger.

17. The egg/larva/pupa/adult of *Tetramorium caespitum* L.

7. Insectivorous culture as sustainable development of food in China

China has a long insectivorous culture and important insect food resources. The characteristics of insect diversity and high nutritional value promise the great potentials to utilize these precious resources. At present, the development and utilization of China's edible insects are mainly in two aspects: the insect dishes and insect protein products.

7.1. Insect dishes in China quickens your appetite

Chinese insect dishes and foods often include traditional and innovative ways of two eating, such as oil silkworm chrysalis, Cicada Fried Crispy Fried locust Lang, Chinese caterpillar fungus duck and Tremella silkworm chrysalis, etc. Also, the insects could be used as ingredients for bread, insect drinks and wine. Edible insects can also be processed into insect nutrients liquid and health products as the main raw materials that are currently fashionable gifts to share for human health. And, most of this kind of products in China are listed as follows: The proteolytic enzyme of honeybee pupa and Bee pupa drinking; Royal Jelly Capsules, Royal jelly cream and Royal jelly powder; Drone pupa wine, Gekko-drone wine which is prepared with Gekko, drone pupa, Lycium chinense and rice wine and through twice soaking. Tussah pupa protein drink for quick nutrition supplement; Amino acid drink of yellow powder insect caterpillar protein; Amino acid drink of sweet potato hawkmoth canned; Soy sauce of silkworm chrysalis; Dry bread cake of silkworm chrysalis protein; Silk protein beverage and jelly, silkworm moth oil and wine, etc. In addition, the raw materials of nutritional health products used more ants as well, which will satisfy the demands of treating different diseases like tracheitis, active chronic gastritis, dysmenorrhea, psychoneurosis, pulmonary tuberculosis, alopeci and impotence, etc.

7.2. The development of insect protein products in China

The fly pupae are mainly used as raw material to develop traditional curative food and make the high protein food through processing. These research works currently focus on the development and utilization of several kinds of insects: flies, locust, cicadas and silkworms and so on. Then, this kind of eating insects is rich in protein, which can be compared to other foods such as meat and eggs. They can, not only solve the food shortage, but also food crisis of global food inequality in the near future.

Extensive and profound Chinese food culture is glamorous. The miracle of insectivorous culture refers to a gorgeous art of diet culture treasure in China. At once, as one of the most urgent tasks in the development and utilization of insects, edible insects can be converted to a steady stream of elegant food dedicated to all humanity.

Author details

Weiping Yin*, Junna Liu, Huaqing Liu and Biyu Lv

*Address all correspondence to: yinwp@haust.edu.cn

School of Chemical Engineering and Pharmaceutics, Henan University of Science and Technology, Luoyang, PR China

References

[1] José M, Murosa S, Barroso FG, et al. Insect meal as renewable source of food for animal feeding: A review. Journal of Cleaner Production. 2013;**4**:1-43

[2] Defoliart GR. Insects as food: Why the western attitude is important. Annual Review of Entomology. 1999;**44**:21-50

[3] Zhou ZH, Yang W, Xu DP, et al. Analysis of resource value of edible insect. Science and Technology of Food Industry. 2014;**17**:390-394 (in Chinese)

[4] Bodenheimer FS. Insects as Human Food. The Hague, Netherlands: Junk Publishers; 1951

[5] Gullan PJ, Cranston PS. The Insects: An Outline of Entomology. Oxford: Blackwell Publishing Ltd; 2004. pp. 1-584

[6] Huis AV, Itterbeeck JV, Klunder H, et al. Edible Insects: Future Prospects for Food and Feed Security. Rome: Food and Agriculture Organization of the United Nations; 2013. pp. 67-80

[7] Klunder HC, Rooijackers WJ, Korpela JM, et al. Microbiological aspects of processing and storage of insects. Food Control. 2012;**26**:628-631

[8] Shockley M, Dossey AT. Mass Production of Beneficial Organisms' Invertebrates and Entomopathogens. USA: Academic Press; 2014. pp. 617-652

[9] Rumpold BA, Schlüter OK. Potential and challenges of insects as an innovative source for food and feed production. Innovative Food Science & Emerging Technologies. 2013;**17**:1-11

[10] Deng ZB, Yang W, Yang CP, et al. Nutrition analysis and evaluation from wasps. Acta Nutrimenta Sinica. 2013;**35**(5):514-515 (in Chinese)

[11] Wen LZ. Nutrition of edible insects in Mexico. Entomological Knowledge. 1998;**35**(1):58-61 (in Chinese)

[12] Huang Q, Zhou ZJ, Zhou DG, et al. Nutritional analysis of seven species of insects. Acta Nutrimenta Sinica. 2007;**29**(1):94-96 (in Chinese)

[13] DeFoliart GR. Encyclopedia of Insects. 2nd ed. USA: Academic Press; 2009. pp. 376-381

[14] Feng Y, Chen XM, Wang SY, et al. Common edible insects and nutritional value from Hemiptera. Forest Research. 2000;**13**:608-612 (in Chinese)

[15] Barker D, Fitzpatrick MP, Dierenfeld ES. Nutrient composition of selected whole. Invertebrates Zoo Biology. 1998;**17**:123-134

[16] Oliveira JS, de Carvalho JP, de Sousa RF, et al. The nutritional value of four species of insets consumed in Angola. Ecology of Food and Nutrition. 1976;**5**:91-97

[17] Rong BX, Gan SY. Ants and preparation of trace element analysis. Chinese Traditional and Herbal Drugs. 1987;**18**:47-49 (in Chinese)

[18] Jiang SJ. Chinese Medicinal Insects' Integration. Beijing: China Forestry Publishing House; 1999. pp. 9-13 (in Chinese)

[19] Wen LZ. Edible Entomology Principle and Application. Changsha: Hunan Science and Technology Press; 1998. pp. 71-146 (in Chinese)

[20] Liu XG, Ju XR, Wang HF, et al. Insect oil and its nourishment appraisement. Journal of the Chinese Cereals and Oils Association. 2003;**6**:11-13

[21] Xu JP. Food Nutrition and Health. Hefei: University of Science and Technology of China Press; 2002

[22] Matthew G, Schroeder FC. Insect Natural Products; Comprehensive Natural Products II, Elsevier Ltd, Oxford, 2010, pp. 67-103

[23] Chen XM, Feng Y. China's Edible Insects. Beijing: China Science and Technology Press; 1999

Chapter 3

Selected Edible Insects and Their Products in Traditional Medicine, Food and Pharmaceutical Industries in Africa: Utilisation and Prospects

Afam I.O. Jideani and Rinah K. Netshiheni

Additional information is available at the end of the chapter

http://dx.doi.org/10.5772/intechopen.68330

Abstract

Edible insects are a widely exploited food source by many indigenous populations in most regions of the world. Edible insects have been used as food in sub-Saharan Africa, being a good source of protein and essential amino acid. Historically, they are important contributors to improving food and nutrition security, particularly for people who suffer from malnutrition due to protein deficiency. Africa is the continent with the highest number of records (19 countries), followed by the USA (5 countries) and Asia (5 countries). The species used for food and as feed include *Hodotermitidae, Kalotermitidae, Rhinotermitidae* and *Termitidae*. Insects, such as termites, are also eaten raw directly on emergence from the holes. Species used in traditional popular medicine include *Hodotermitidae* and *Termitidae*. They are used in the treatment of various diseases that affect humans such as influenza, asthma, bronchitis, whooping cough, sinusitis, tonsillitis and hoarseness.

Keywords: termites, *Isoptera*, micronutrients, therapeutics, utilisation, food security

1. Introduction

1.1. Overview on edible insects in sub-Saharan Africa

Insects have been consumed for generations in many regions of the world, a practice that has increased in popularity in recent years [1–6]. More than 2.5 billion people, mainly in Africa and Asia, commonly consume insects [7]. The use of termites was registered in 29 countries over three continents. Africa is the continent with the highest number of records (19 countries), followed by the USA (5 countries) and Asia (5 countries). Throughout the world, a large portion of the human population consumes insects as a regular part of their diet. Insects' consumption, called Entomophagy, has played an important role in the history of human nutrition in Africa [8]. Currently, edible insects are gaining much attention for their high nutritional value and environmental advantages over meat production [9]. Their short life cycles, low space requirements, efficient nutrient conversion rates and lower greenhouse gas production render insects to be, in principle, an excellent alternative to meat [5]. It is postulated that termites contain high-quality nutrients including highly digestible proteins [10], as well as minerals, which are more bioavailable than minerals from plant foods [11]. Crickets in particular are believed to have higher quality animal protein than some conventional sources, such as fish, and are more affordable among poor communities [12]. Currently, attention is being drawn to this valuable traditional food resource, which if tapped or exploited could contribute to a more sustainable solution for malnutrition in sub-Saharan Africa. It is estimated that insects form part of the traditional diets of at least two billion people. Thousands of edible species have been identified [13–15]. More than 1900 species have reportedly been used as food [16, 17]. Insects deliver a host of ecological services that are fundamental to human survival. Termites are eusocial insects belonging to the family *Isoptera* that play a major role in the tropical ecosystem (**Figure 1**).

In general, insects are important in plant reproduction, waste bio-conversion and in biocontrol of harmful pest species leading to a variety of valuable food and non-food products used in applications such as maggot therapy [1]. They are used as collection items and ornaments and in movies, visual arts and literature. Globally, the most commonly consumed insects are beetles (*Coleoptera*, 31%), caterpillars (*Lepidoptera*, 18%) and bees, wasps and ants (*Hymenoptera*, 14%), grasshoppers, locusts and crickets (*Orthoptera*, 13%), cicadas, leafhoppers, planthoppers, scale insects and true bugs (*Hemiptera*, 10%), termites (*Isoptera*, 3%), dragonflies (*Odonata*, 3%), flies (Diptera, 2%) and other orders (5%) (**Figure 2**).

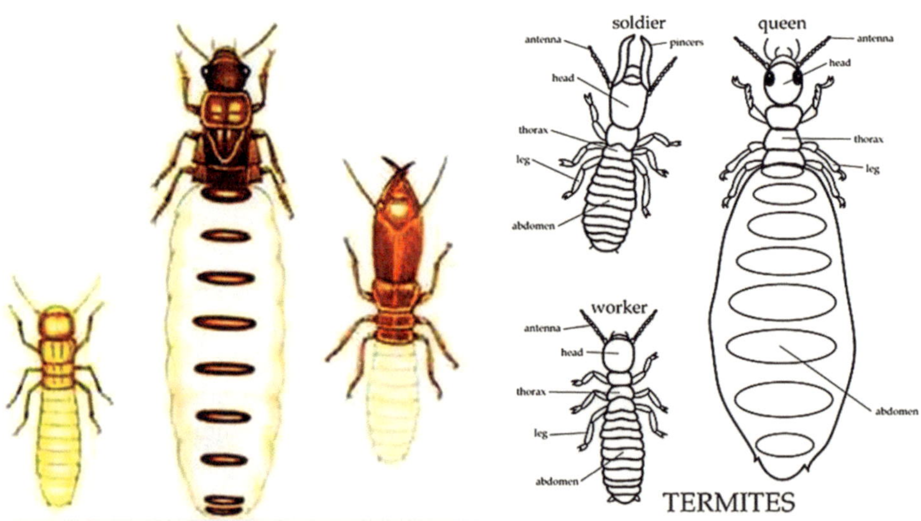

Figure 1. Structure of edible termites [18].

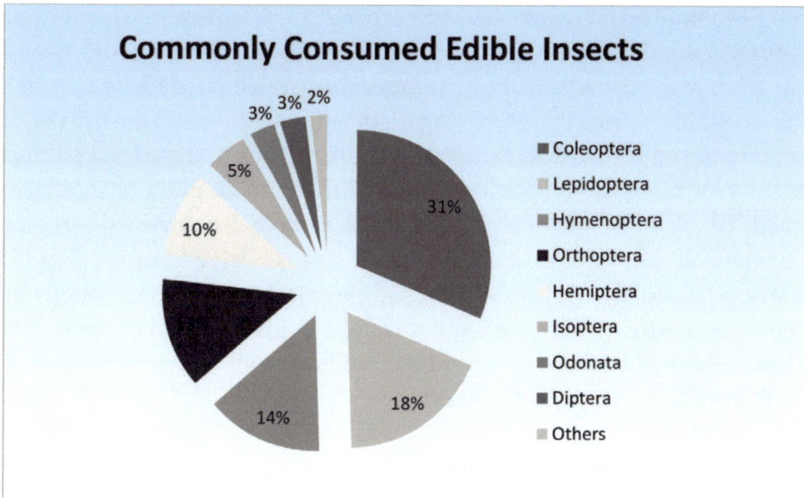

Figure 2. Commonly consumed edible insects globally [17].

The importance of insects as a food source for human is not surprising, since this is the group with the highest number of species in nature, thereby representing significant biomass [19]. Considered as important natural resources, insects are in many ways a basic component of the diets of humans and other animals [20] and have played an important role as a source of medicinal resources [21–24].

By 2050, the global population is estimated to be nine billion people and could possibly lead to a global food demand increase of up to 70% compared with our current food requirements [25]. Invariably, conventional sources of protein may not be sufficient for the global human population; hence, alternative sources such as insects will be required.

2. Edible insects in sub-Saharan Africa

2.1. Cultivation

The cultivation of edible insects in sub-Saharan Africa is not common as obtainable in developed economies. Although edible insects are popular in the continent, they are not cultivated and there appears to be no campaign strategies in place. This is attributable to a number of reasons ranging from cultural, behavioural to lack of understanding of nutritional benefits of insects in the continent. Rearing insects may also require minimal land or market introduction efforts, as insects already form part of some local food cultures. A recent study shows that house crickets (*Acheta domesticus*), common across Africa, are a highly valuable yet neglected source of proteins [12] contributing a viable solution to food security challenges.

Apparently, due to huge local and international demand for insect-based foods, insect farming initiatives seems to have taken off in Africa. Iwuoha [26] highlighted the following initiatives:

(i) The McGill Entomophagy Project—in September 2013, the Clinton Global Initiative seems to have awarded a significant prize to a team of student entrepreneurs from the McGill University in Montreal, Canada to fight hunger and nutrition deficiencies in developing regions of the world like Africa by improving diets with insect-based meals. This team, working with researchers and local insect farmers in Thailand, has assessed red palm weevil larva farming practices for possible translation to West Africa where palm weevil larva are eaten, but not farmed; (ii) Prof. Arnold van Huis (Insect Protein for Africa)—a researcher using insects as a sustainable source of protein for Africa—is on a drive to raise 1.5 million Euros in funding to carry out fundamental research of simple insect cultivation methods in Africa and (iii) AgriProtein—a South African business developing insect-based protein feed, extruded oil, and fertilisers since 2009 and has raised $11 million from strategic partners to commercialise and globalise its revolutionary concept.

2.2. Cultural and religious considerations

The consumption of insects as food is acceptable and practised by many cultures around the world [18]. Many ethnic groups in 130 countries utilise insects as essential elements of their diet, even in the United States there has been an increasing interest in insect-based food products in recent years [27]. Edible insects are a widely exploited food source by many indigenous populations in most regions of the world [18].

Entomophagy, the consumption of insects, is heavily influenced by cultural and religious practices in many regions of the world [28] and sometimes economic circumstance in some parts of Africa. Historical references to the use of insects for food are currently found in some religious literatures, for example, in the Bible, in Leviticus chapter 11, verse 22 and in Matthew chapter 3, verse 4 [29]. Despite these references, the topic of entomophagy has only very recently started to capture some attention worldwide. In the most developed economy, however, people still view entomophagy with disgust and associate eating insects with primitive behaviour. This attitude has resulted in the neglect of insects in agricultural research.

Edible termites form an important part of the food culture in the Lake Victoria region of East Africa [30, 31]. The study of Pambo et al. [32] shows that in Kenya attitudes, subjective norms and behavioural capabilities are considerations in designing food from edible insects (FEI) that are culturally appropriate. Their study proposed FEI as a viable option, given that insects are ubiquitous and that taste, availability/convenience, cost, nutrition, health benefits/risks and disgust factors are major determinants of consumption. Their study was conceptualised to identify potential differences in salient beliefs between participants from western region and those from eastern region, given that consumption of edible insects is popular in the western region. Although Pambo et al. [32] did not find potential differences in salient beliefs regarding insect consumption between two regions in Kenya, differences could exist between countries. Acceptance and popularity of FEI would depend much on campaign, nutrition literacy and other considerations. Kipkoech [12] observed that it took the presence of participants from different countries, continents and cultures, some of whom had tasted cricket before to demystify their consumption as food at a conference reception.

2.3. Consumption

Insects, especially termites, have been used as food in sub-Saharan Africa, Asia, Australia and Latin America (**Table 1**). Edible insects consumption has been recently reported in Nigeria [33, 34] among which the termites had the highest mean frequency [33].

Some of the insects are eaten raw directly from the emergence hole [35, 36]. Although termite harvest begins with the onset of the rains and the swarming of the winged termites, villagers have shown that some termites could be induced to emerge even during the dry seasons,

Species	Country (ies)	Reference
Hodotermitidae		
Hodotermes mossambicus	Botswana	Gahukar [41]
Microhodotermes viator	South Africa	DeFoliart [18]
Kalotermitidae		
Kalotermes flavicollis	Brazil, Thailand	Jongema [17]
Rhinotermitidae		
Coptotermes formosanus	China	Jongema [17]
Reticulitermes flavipes	Thailand	Wilsanand [2]
Reticulitermes tibialis	Mexico	DeFoliart [18]
Termitidae		
Cubitermes atrox	Indonesia	Jongema [17]
Labiotermes labralis	Columbia	Jongema [17]
Macrotermes acrocephalus	China	Jongema [17]
Macrotermes falciger	Zimbabwe, South Africa	Wilsanand [2]
Macrotermes gabonensis	Congo	Jongema [17]
Macrotermes herus	Tanzania	McGrew and Roger [42]
Macrotermes lilljeborgi	Cameroon, Guinea	Deblauwe and Janssens [43]
Macrotermes michaelseni	Malawi	Sileshi et al. [44]
Macrotermes sub hyalinus	Angola, Zambia, Kenya	Lesnik [45]
Macrotermes vitrialatus	Zambia	Jongema [17]
Microcerotermes dubius	Malaysia	Jongema [17]
Nasutitermes ephratae	Venezuela	Jongema [17]
Odontotermes badius	South Africa, Zambia	Jongema [17]
Odontotermes capensis	South Africa	Jongema [17]
Odontotermes kibarensis	Uganda	Sileshi et al. [44]

Table 1. Termite species used as food or feed.

making them available throughout the year. Banjo et al. [37] were of the view that this has created attachment to the termite enterprise by locals to the extent that in some parts of the region, termite mounds are owned by individuals and sometimes form part of inheritance when one dies.

In many households, termites are a delicacy enjoyed by almost all ethnic communities in western Kenya. They are consumed as part of a meal or as a complete meal with tapioca, bread, roast corn or simply eaten as snack food. Some mothers grind the dried termites into flour and use it as a sprinkle in baby porridge [38]. Termites are also eaten raw directly from the emergence hole [35, 36]. Addition of termites to maize resulted in significant increase in β-carotene, niacin, vitamin B6, and B12 content, with significant reduction in thiamine, riboflavin and ascorbic acid content of enriched complementary foods.

The utilisation of insects as a sustainable and secure source of animal-based food for the human diet has continued to increase in popularity in recent years [39]. Many ethnic groups in approximately 130 countries utilise insects as essential elements of their diet. In recent years, in the United States, there has been an increasing interest in insect-based food products [27]. More than 2000 insect species and other small invertebrates are consumed as food by humans and animals alike [40] and used for either self-sufficiency or commercial food products in many parts of the world.

Insects link biodiversity conservation and human nutrition in a way that many other food sources do not. They often contain more protein, fat and carbohydrates than equal amounts of beef or fish, and a higher energy value than soy beans, maize, beef, fish, lentils or other beans [46].

3. Processing of edible insects

Insects are often consumed whole but can also be processed into granular or paste forms. In most western and eastern African countries, termites are collected during the rainy seasons as they emerge from holes on the ground. In western Kenya, this period is between April and October. They are prepared by blanching in boiling water, dried in the sun before frying in their own fat. Others dipped the crickets into hot water for 1 min and then sun-dried and ground them for use in making porridge, cookies and other sweet delicacies or deep-fry to get crispy crickets that would be eaten whole [12]. Surprisingly, the deep-fried crickets were everyone's favourite because of their delicious aroma and taste.

Several studies have shown that the product preparation affects the willingness to eat insects [47]. However, to date, consumer expectations and preferences towards different aspects of the product preparation have not been investigated, especially in sub-Saharan Africa. **Figure 3** shows the processing of mopane worms by frying.

Roasted termites moisture content was very low, while the crude protein, fat, carbohydrate and gross energy content were very high. Banjo et al. [37] stated that there was a significant reduction in moisture content of formulated complementary foods, the level of reduction increasing with increasing level of 10–20% inclusion of termites ($p<0.05$) with the values increasing with

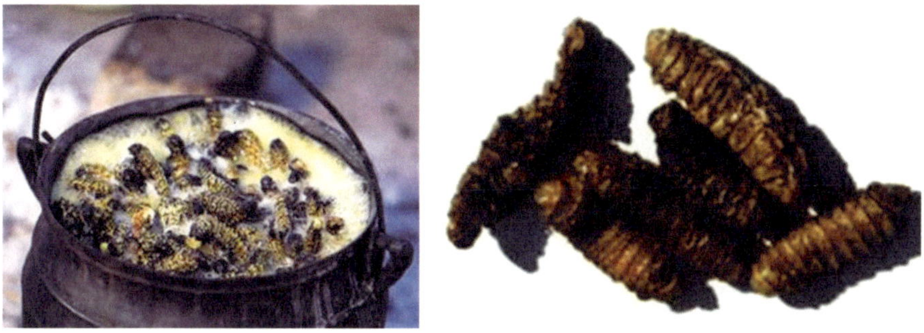

Figure 3. Mopane worm. http://int.search.myway.com/search/AJimage.jhtml [03 December 2016].

increasing level of inclusion of termites. They also reported significant increase in values of crude protein and fat, ash, total carbohydrates, mineral and gross energy of the formulated maize and sorghum complementary foods ($p<0.05$). However, they highlighted that some of the increases in value were lower than recommended by FAO/WHO.

4. Edible insects and bio-economy

Currently, attention is being drawn to this valuable traditional food resource, which if tapped or exploited is likely to be a more sustainable solution for nutrient deficiency. Considering the trajectory of the economy in sub-Saharan Africa, the gathering and farming of insects can offer employment and cash income, either at the household level or in larger, industrial-scale operations. In most parts of sub-Saharan Africa and Southeast Asia, the process of insect gathering, rearing and processing into street foods or for sale as chicken and fish feed is easily within reach of small-scale enterprises. With only a few exceptions, international trade in edible insects is insignificant, but border trade is significant. The trade that does exist in developed countries is often driven by demand from immigrant communities or because of the development of niche markets that sell exotic foods.

Some of the poorest people in Africa, living in urban and rural areas, can earn a living by the gathering, cultivation, processing and sale of edible insects. Such activities can improve diets, making available cash income by selling raw and processed insects as street foods or supply junior and high schools where school meals are served to pupils. Insects can be directly and easily collected from nature or farmed with minimal technical or capital expenditure considering the use of basic rearing and harvesting equipment.

5. Nutraceutical benefits

The nutraceutial benefit of insects is substantial varying as a result of the wide range of edible species. Within the same species, it has been shown that nutritional and medicinal values could differ based on the metamorphic stage of the insect, the habitat, and diet. They are a readily available source of protein (**Figure 4**), lipids, carbohydrates, certain vitamins, and minerals such as calcium, iron, or zinc (**Table 2**). The protein content in insects is high, equivalent to that of fish and meat and is also said to be similar to the one found in a human body, making it easier to be utilised by the body as to compared to plant protein [13]. The energy content of insects is on average comparable to that of meat (on a fresh weight basis) except for pork because of its particularly high-fat content [48].

Dried termites have been reported to be a good source of dietary protein, fat and micronutrients [37, 50–52, 34]. Adepoju and Omotayo [34] reported termites to be low in anti-nutrients and suggested its possible inclusion in formulating adequate, nutrient-dense complementary foods with nutraceutical benefits. Addition of termites to maize resulted in significant increase in β-carotene, niacin, vitamin B6 and B12 content, with significant reduction ($p<0.05$) in thiamine, riboflavin and ascorbic acid content [37].

Termites constitute a food source of great nutritional value: high in protein and essential amino acids such as tryptophan, which is generally limiting in the food insects [53]. Termites are rich in minerals and other micronutrients. Essential fatty acids are well represented [54]. In general, heads of termites are better nutritionally featured than thorax and abdomens. Termites are undeniably rich sources of iron and their inclusion in the daily diet could improve iron status and help prevent anaemia in developing countries. Essential vitamins for stimulating metabolic processes and enhancing immune system functions are present in most edible insects such as termites. Booth [54] showed for a whole range of insects that thiamine ranged from 0.1 to 4 mg/100 g of dry matter. Insects contain significant amounts of fibre, as measured by crude fibre, acid detergent fibre and neutral detergent fibre. The most common form of fibre in insects is chitin, an insoluble fibre derived from the exoskeleton [17].

From a utilitarian perspective, termites are commonly used insects in entomotherapeutic practices and traditional popular medicine [20, 23, 55, 56] in the treatment of various diseases (**Table 3**) that affect humans, such as influenza, asthma, bronchitis, whooping cough, sinusitis, tonsillitis and hoarseness [57, 58]. They have historically been an important source of food that may contribute to improving human diet, particularly for people who suffer from malnutrition due to a deficit of protein [17], as they are considered a nonconventional food with great economic and social importance [3–6].

Figure 4. Some edible insects [49].

Insect or food item	Protein (g/kg)	Fat (g/kg)	Calories (kcal/kg)	Thiamine (mg/kg)	Riboflavin (mg/kg)
Black soldier fly	175	140	1994	7.7	16.2
House fly	197	19	918	13.3	77.2
House cricket	205	68	1402	0.4	34.1
Super worm	197	177	2423	0.6	7.5
Meal worm	187	134	2056	2.4	8.1
Giant mealworm	184	168	2252	1.2	16.1
Wax worm	141	249	2747	2.3	7.3
Silk worm	93	14	674	2.3	9.4
Beef	256	187	2776	0.5	1.8
Powder milk	165	268	4982	2.6	14.8

Source: Finke [50].

Table 2. Nutritional content of insects compared with other high-protein foods.

Species/family	Treated disease	Country	Reference
Hodotermitidae			
Hodotermes mossambicus	Child malnutrition	Zambia (Africa)	Cheng and Feng [59]
Termitidae			
Macrotermes bellicosus	Suture wounds	Somalia (Africa)	Wilsanand et al. [60]
Macrotermes nigeriensis	Wounds	Nigeria (Africa)	Alves [61]
Macrotermes exiguus	Asthma, flu etc.	Brazil	Alves [23]
Odontotermes feae	Ulcer, Rheumatics	India	Solavan et al. [55]
Pseudacanthotermes spinger	Antifungal properties	Brazil	Countinho [56]

Table 3. Termite species used in traditional medicine.

6. Prospects and challenges of edible insects industry

The induction of insect emergence during off seasons by villages [37] gives room for technological application in making insects available all through the year. This is an area that needs investigation as the world looks for alternative, environmentally friendly and cheaper sources of protein foods.

As earlier highlighted, edible termites are a good source of protein and the essential amino acid tryptophan. Many commercial food products are enriched with protein extracted or derived from legumes but insect protein is better in terms of nutritional properties and contains all the essential amino acids (**Table 2**). Adepoju and Omotayo [34] reported termites as being low in anti-nutrients and suggested their possible inclusion in formulating adequate, nutrient-dense complementary foods. Moreover, some insects (**Table 4**) are richer in protein

Insect order	Stage	Range (% protein)
Coleoptera	Adult and larvae	23–66
Lepidoptera	Pupae and larvae	14–68
Hemiptera	Adult and larvae	42–74
Homoptera	Adult, larvae and egg	45–57
Hymenoptera	Adult, pupae, larvae and egg	13–177
Odonata	Adult and naiad	46–65
Orthoptera	Adult and nymph	23–65

Source: Xiaoming et al. [63].

Table 4. Crude protein content of some insect order.

than beans (23.5% of protein), lentils (26.7%) or soybean (41.1%) [62]. At present, extraction processes for specific food ingredients are too costly and will need to be further developed to render insects profitable and applicable for industrial use in the food and feed sectors.

In the continent, the role of the media, peers and health officials in a FEI intervention need to be given consideration [32]. A theory-based intervention targeting FEI consumption and nutrition literacy, and ultimately increase the intake of FEI among the target population was proposed by Pambo et al. [32].

As stated by Pambo et al. [32], population growth, urbanisation and climate change are among the factors that have created uncertainties and pressures on current global food and economic systems. Insects as food and consumption of foods from edible insects are being promoted as one potential solution to the declining access to protein foods. It has been said by various authors that with the increasing global population, one of the strategies to improve food and nutrition security is to diversify diets using available food sources [25, 64, 65, 46, 3, 5, 12]. However, one of the challenges facing consumption [66] of edible insects is the limited information regarding consumer-psychographic characteristics including attitudes, values, interests and beliefs.

7. Conclusion and recommendation

Insects abundantly thrive in Africa, given the prevailing tropical climate. As the price of beef, chicken and fish continue to rise across the world, there is a huge opportunity for insects to meet the animal protein needs of human beings and livestock. It is now the view that edible insects have more iron than sirloin beef. One area of challenge in edible insect promotion, consumption and commercialisation is the consumer-psychographic characteristics. Edible insects are a highly nutritious and healthy food source with high fat, protein, vitamin, fibre and mineral content. The consumption of edible insects should be emphasised as this could be one of the sustainable strategies towards the alleviation of poverty, hunger and malnutrition in sub-Saharan Africa.

Author details

Afam I.O. Jideani* and Rinah K. Netshiheni

*Address all correspondence to: afam.jideani@univen.ac.za

Department of Food Science and Technology, University of Venda, Thohoyandou, South Africa

References

[1] Van-Huis H. Insects as food in sub-Saharan Africa. Insect Science and Its Application. 2003;**23**:163-185

[2] Wilsanand V. Utilization of termite, *Odontotermes formosanus* by tribes of South India. Journal of Medicine and Food. 2005;**4**:121-125

[3] Johnson DV. The contribution of edible forest insects to human nutrition and to forest management: Current status and future potential. In: Durst PB, Johnson DV, Leslie RN, Shono K, editors. Forest Insects as Food: Humans Bite Back. Thailand: Food and Agriculture Organization of the United Nations; 2010. pp. 5-22

[4] Niaba KV, Atchibri LO, Gbassi KG, Beugre AG, Adou M, Anon AB. Consumption survey of edible winged termites in Cote d'Ivoire. International Journal of Agriculture and Food Science. 2012;**2**:149-155

[5] Van Huis A; Vantomme P. Conference report: Insects to feed the world. Food Chain. 2014;**4**:184-193

[6] Shockly M, Dossey AT. Insects for human consumption. In: Morales-Ramos J, Rojas G, Shapiro-Ilan DI, editors. Mass Production of Beneficial Organisms. London: Elsevier; 2014. pp. 617-652

[7] FAO (Food and Agriculture Organization of the United Nations). Biodiversity and sustainable diets: United against hunger. Report presented at World Food Day/World Feed Week; 2-5 November 2010; Rome

[8] Srivastava SK, Babu N, Pandey H. Traditional insect bioprospecting: As human food and medicine. Indian Journal of Traditional Knowledge. 2009;**8**:485-494

[9] Van-Huis H, Van Gurp H, Dicke M. Potential of insects as food and feed in assuring food security. Annual Review on Entomology. 2013;**58**:563-583

[10] Kinyuru JN, Kenji GM, Muhoho SN. Nutritional potential of longhorn grasshopper (*Ruspolia differens*) consumed in Siaya District, Kenya. Journal of Agriculture, Science and Technology. 2010;**12(1)**:1-24

[11] Omotoso OT. Nutritional quality, functional properties and anti-nutrients of the larva of *Cirinaforda* (Westwood) (Lepidoptera: Satuniidae). Journal of Zhejang University. 2006;**7(5)**:51-55

[12] Kipkoech C. Neglected insects as a remedy for food insecurity: The case of crickets. RUFORUM Weekly. 2017;**1**: 3. A weekly digest of News, Events and Opportunities. RUFORUM Communication communications=ruforum.org@mail75.suw11.mcdlv.net

[13] Bukkens SGF. The nutritional value of edible insects. Ecology of Food and Nutrition. 1997;**36**:287-319

[14] Bukkens SGF, Paoletti, MG. Insects in the human diet: Nutritional aspects. In: Paoletti MG, editor. Ecological Implications of Mini Livestock: Role of Rodents, Frogs, Snails, and Insects for Sustainable Development. New Hampshire: Science Publishers; 2005; pp. 545-577

[15] DeFoliart GR. Insects as food: Why the western attitude is important. Annual Review of Entomolology. 1999;**44**:21-50

[16] Ramos-Elorduy J. Energy supplied by edible insects from Mexico and their nutritional and ecological importance. Ecology of Food and Nutrition. 2009;**47**:280-297

[17] Jongema Y. List of edible insect species of the world. 2014; http://www.wageningenur.nl/en/ExpertiseServices/Chairgroups/PlantSciences/Laboratory-of-Entomology/Edible-insects/Worldwidespecies-list.htm. [Accessed from 05 November 2014]

[18] DeFoliart GR. The human use of insects as food resource: A bibliographic account in progress. 2002; http://www.food-insects.com/book7_31/The%20Human%20Use%20of%20Insects%20as%20a%20Food%20Resource.htm]

[19] Meyer-Rochow VB, Chakravorty J. Notes on entomophagy and entomotherapy generally and information on the situation in India in particular. Journal of Ethnobiology and Ethnomedicine. 2013;**48**:105-112

[20] Raubenheimer D, Rothman J. Nutritional ecology of entomophagy in humans and other primates. Annual Review of Entomology. 2012;**58**:141-160

[21] Costa-Neto EM. Entomotherapy, or the medicinal use of insects. Journal of Ethnobiology. 2005;**25**:93-114

[22] Dossey AT. Why insects should be in your diet. Scientist. 2010;**27**:22-23

[23] Alves RRN. Fauna used in popular medicine in Northeast Brazil. Journal of Ethnobiology and Ethnomedicine. 2011;**5**:1-30

[24] Alves RRN, Albuquerque UP. Animals as a source of drugs: Bioprospecting and biodiversity conservation. In: Alves RRN, Rosa IL, editors. Animals in Traditional Folk Medicine: Implications for Conservation. Heidelberg: Springer; 2013. pp. 67-89

[25] FAO (Food and Agriculture Organization of the United Nations). How to Feed the World in 2050. 2009. http://www.fao.org/fileadmin/templates/wsfs/docs/expert_paper/How_to_Feed_the_World_in_2050.pdf [Accessed 10 March 2015]

[26] Iwuoha J-P. The Top 4 Most Widely Eaten Insects in Africa, and Why Insect Farming Has Become a Big Business Opportunity. 2015. http://www.smallstarter.com/browse-ideas/top-4-most-eaten-insects-in-africa-and-why-insect-farming-is-a-huge-business-opportunity/Agribusiness and Food, Business ideas. Tweets by @smallstarters. [Accessed 22 January 2017]

[27] Dossey A, Morales-Ramos J, Guadalupe R. Insects as Sustainable Food Ingredients: Production, Processing and Food Applications. 1st ed., 2016. Academic Press, Cambridge, MA, USA.

[28] Cerritos R. Insects as food: An ecological, social and economical approach. CAB Reviews: Perspectives in Agriculture, Veterinary Science, Nutrition and Natural Resources. 2009;**4(27)**:1-10

[29] James K. The Holy Bible. Authorised King James Version (KJV). World Bible Publishers, Inc.: USA; 1986

[30] DeFoliart GR. Insects as human food: Some nutritional and economic aspect. Crop Protection. 1999;**11**(5):395-599

[31] Ayieko MA. Nutritional value of selected species of reproductive *Isoptera* and Ephemeroptera within the ASAL of Lake Victoria basin. Journal of Discovery and Innovation. 2007;**19(2)**:126-130

[32] Pambo KO, Mbeche RM, Okello JJ, Kinyuru JN, Mose GN. Consumers' salient beliefs regarding foods from edible insects in Kenya: A qualitative study using concepts from the theory of planned behaviour. African Journal of Food, Agriculture, Nutrition and Development. 2016;**16(4):**11366-11385

[33] Cloutier J, editor. Edible Insects in Africa: An Introduction to Finding, Using and Eating Insects. 1st ed. Wageningen Digigrafi, Veenendaal, Netherland: Agromisa Foundation and CTA; 2015. ISBN Agromisa: 978-90-8573-146-7, ISBN CTA: 978-92-9081-577-8

[34] Adepoju OT, Omotayo OA. Nutrient composition and potential contribution of winged termite (*Marcrotermes bellicosus* Smeathman) to micronutrient intake of consumers in Nigeria. British Journal of Applied Science and Technology. 2014;**4(7)**:1149-1158

[35] Christensen DL, Orech FO, Mungai MN, Larsen T, Friss H, Aagaard-Hansen J. *Entomophagy* among the Luo of Kenya: A potential mineral source? International Journal of Food Science and Nutrition. 2006;**57:**198-203

[36] Ayieko MA., Ndong'a FO, Tamale A. Climate change and the abundance of edible insects in the Lake Victoria Region. Journal of Cell and Animal Biology. 2010;**7**:112-118

[37] Banjo AD, Lawal OA, Songonuga EA. The nutritional value of fourteen species of edible insects in southwestern Nigeria. African Journal of Biotechnology. 2006;**5**:298-301

[38] Bergeron D, Rodney JB, Franklin LR, Irv K, John O, Alfred AB. The nutrient composition of an insect flour sample from Lake Victoria, *Uganda*. Journal of Food Composition and Analysis. 1988;**11**:371-377

[39] Dossey AT. Insects and their chemical weaponry: New potential for drug discovery. 2013;**27**:1737-1757

[40] de Figueirêdo RECR, Vasconcellos A, Policarpo IS, Alves RRN. Edible and medicinal termites: A global overview. Journal of Ethnobiology and Ethnomedicine. 2015;**11**:29. DOI: 10.1186/s13002-015-0016-4

[41] Gahukar RT. Food security: The challenges of climate change and bioenergy. Current Science. 2009;**96**:26-28

[42] McGrew WC, Roger E. Brief report: Chimpanzees, tools, and termites: New record from Gabon. American Journal of Primatology. 1985;**5**:171-174

[43] Deblauwe I, Janssens G. New insights in insect prey choice by chimpanzees and gorillas in southeast Cameroon: The role of nutritional value. American Journal of Physics and Anthropology. 2008;**135**:42-55

[44] Sileshi GW, Nyeko P, Nkunika POY, Sekematte BM, Akinnifesi FK, Ajayi OC. Integrating ethno-ecological and scientific knowledge of termites for sustainable termite management and human welfare in Africa. Ecology. 2009;**14**:1-21

[45] Lesnik JL. Termites in the hominin diet: A meta-analysis of termite genera, species and castes as a dietary supplement for South African robust australopithecines. Journal of Human Evolution. 2014;**71**:1-11

[46] FAO. Edible Insects: Future Prospects for Food and Feed Security. Forestry Paper No. 171. Rome: Food and Agriculture Organization of the United Nations; 2013

[47] Tan JC, Jakob WB, Rossow G. Increase in tropical rainfall driven by changes in frequency of organized deep convention. Nature. 2015;**519**:451-454

[48] Sirimungkararat S, Saksirirat W, Nopparat T, Natongkham A. Edible products from eri and mulberry silkworms in Thailand. In: Durst PB, Johnson DV Leslie RN, Shono K, editors. Forest Insects as Food: Humans Bite Back. Bangkok, Thailand: FAO; 2010. pp. 189-200

[49] NEWS24. Top 5 insects you should be eating. 2017. http://www.news24.com/Green/News/Top-5-insects-you-should-be-eating-20140708

[50] Finke MD. Complete nutrient content of four species of feeder insects. Zoology Biology. 2012, 32(1): 27-36.

[51] Ekpo KE, Onigbinde AO, Asia IO. Pharmaceutical potentials of the oils of some popular insects consumed in Southern Nigeria. African Journal of Pharmacy and Pharmacology. 2009;**3**:51-57

[52] Adeyeye EI. Fatty acid composition of *Zonocerus variegatus, Macrotermes bellicosu*s and *Anacardium occidentale* kernel. International Journal of Pharmacology and Bio-Sciences. 2011;**2(1)**:B135-B144

[53] De Foliart GR. Overview of role of edible insects in preserving biodiversity. In: Paoletti MG, editor. Ecological Implications of Minilivestock. USA: Enfield NH Science Pub; 2005; pp. 123-140

[54] Booth RG. A review of the species resembling *Chilocorus nigrita* (*Coleoptera: coccinellidae*): Potential agents for biological control. Bulletin of Entomological Research. 1998;**88**:361-367.

[55] Solavan A, Paulmurugan R, Wilsanand V. Effect of the subterranean termite used in the South Indian folk medicine. Indian Journal of Traditional Knowledge. 2006;**5**:376-379.

[56] Coutinho HDM, Vasconcellos A, Lima MA, Almeida-Filho GG, Alves RRN. Termite usage associated with antibiotic therapy: Enhancement of aminoglycoside antibiotic activity by natural products of *Nasutitermes corniger* (Motschulsky 1855). BMC Complement Alternative Medicine. 2009;**9**(35):1-4

[57] Alves RRN, Alves HN. The faunal drugstore: Animal-based remedies used in traditional medicines in Latin America. Journal of Ethnobiology and Ethnomedicine. 2010;**7**:1-43

[58] Alves RR, Dias TL. Usos de invertebrados na medicina popular no Brasil esuas implicações para conservação. Tropical Conservation Science. 2010;**5**:159-174

[59] Cheng X, Feng Y. The Edible Insects of Chin. Beijing: Science and Technology Publishing House; 1999; pp. 15-20

[60] Wilsanand V, Preema V, Rajitha P. Therapeutics of insects and insects' products in South Indian traditional medicine. Indian Journal of Pharmacology. 2007;**6**:563-568

[61] Alves RRN. The faunal drugstore: Animal-based remedies used in traditional medicines in Latin America. Journal of Ethnobiology and Ethnomedicine. 2010;**7**:1-43

[62] Ramos-Elorduy J, Moreno JMP, Camacho VHM. Could grasshoppers be a nutritive meal. Food and Nutrition Sciences. 2012;**3**:164-175

[63] Xiaoming C, Ying F, Hong Z, Zhiyong C. Review of the nutritive value of edible insects. In: Durst PB, Johnson DV, Leslie RL, Shono K, editors. Forest Insects as Food: Humans Bite Back. Proceedings of a Workshop on Asia-Pacific Resources and Their Potential for Development; Bangkok: FAO Regional Office for Asia and the Pacific; 2010.

[64] FAO. Forest insects as food: Humans bite back. In: Durst PD, Johnson, DV, Leslie RN, Shona K, editors. In: Proceedings of a Workshop on Asia-Pacific Resources and their Potential for Development; 19-21 February 2008; Bangkok, Thailand: Chiang Mai University

[65] Münke-Svendsen C, Ekesi S, Ayieko M, Kinyuru J, Halloran A, Makkar H and Roos N. Insects as food and feed in Kenya: Past, current and future perspectives. Greeinsect Technical Brief No. 1, Copenhagen, Denmark; 2016.

[66] Verbeke W. Profiling consumers who are ready to adopt Insects as a meat substitute in a western society. Food Quality and Preference. 2015;**39**:147-155.

Chapter 4

Allergy to Edible Insects: A Computational Identification of the IgE-Binding Cross-Reacting Allergen Repertoire of Edible Insects

Pierre Rougé and Annick Barre

Additional information is available at the end of the chapter

http://dx.doi.org/10.5772/68124

Abstract

Allergic manifestations to the ingestion of edible insects have been reported, especially in countries where edible insects are traditionally consumed. However, to date, allergens of edible insects have been poorly investigated. The AllergenOnline server was used for assessing the allergenic character of the putative IgE-binding cross-reactive allergens from the consumed yellow mealworm, silkworm, house fly maggot, migratory locust, house cricket, greater wax moth, black soldier fly, American grasshopper and Indian mealmoth. Positive hits correspond to allergens exhibiting >35% identity over an 80-residue sliding window and 100% identity over an 8-residue sliding window, respectively. Most of the hits consist of allergens from arthropods such as dust mites, crustaceans and insects, and more rarely, of allergens from mollusks, nematodes, and fungi. All the identified allergens share conserved amino acid sequences and three-dimensional structures. Accordingly, the allergens of edible insects form clusters closely related to crustacean, mollusk and nematode clusters into the phylogenetic trees built up from the sequence alignments. Our computational investigations suggest edible insects possess a large repertoire of IgE-binding allergens they share with phylogenetically related groups of arthropods, mollusks, and nematodes. These cross-reacting allergens are susceptible to trigger allergic reactions in individuals previously sensitized to shellfish or mollusks.

Keywords: allergen repertoire, edible insects, shrimps, dust mites, mollusks, IgE-binding cross-reactivity

1. Introduction

The rapidly expanding world population, which is estimated to reach 9 billion people on 2040, underlines the urgent need to develop new sources of food proteins as a complement for the traditionally consumed proteins of plant and animal origin [1–3]. Among the new sources of food and feed proteins, insect proteins appear as a valuable candidate with respect to their good nutritional value for humans and animals [4] and their ability to be produced at a very large industrial scale [5]. However, insect proteins have to be checked for food and feed security before the launching of any large-scale production [6–9]. In this respect, both the chemical (heavy metal and pesticide contamination) and biological safeties including the potential parasitic microbial and parasitic load, and the potential allergenicity, should be evaluated. Depending on the forthcoming predictable introduction of edible insect proteins in both the human and cattle diets, the potential allergic risk associated to the consumption of edible insects has been stressed out, due to the occurrence in insects of pan-allergens common to arthropods, mollusks, and nematodes [8–10]. To date, however, our knowledge on the diversity of insect allergens remains too limited to properly address the potential allergic risk associated to entomophagy, especially for people living in European countries where insect consumption is not a part of their eating habits. In the present chapter, we report the results of a bioinformatic approach aimed at filling the gaps in our existing knowledge about the variety of allergens occurring in some edible insect species.

2. Assessing the complexity of the IgE-binding allergen repertoire of edible insects

A bioinformatic approach based on the AllergenOnline server (http://allergenonline.org) was used for assessing the allergenic character of the putative IgE-binding cross-reactive allergens of some edible insects. Analysis of the available amino acid sequences of putative allergens from yellow mealworm (*Tenebrio molitor*), silkworm (*Bombyx mori*), house fly maggot (*Musca domestica*), migratory locust (*Locusta migratoria*), house cricket (*Acheta domesticus*), grater wax moth (*Galleria mellonella*), black soldier fly (*Hermetia illucens*), American grasshopper (*Schistocerca americana*), and Indian mealmoth (*Plodia interpunctella*) was performed using two large (80 amino acid residues) and restricted (8 amino acid residues) sliding windows, respectively. Positive hits from the AllergenOnline data bank correspond to allergens exhibiting >35% identity over an 80-residue window and 100% identity over an 8-residue window, respectively. The FASTA search algorithm (FASTA 35.04, 2009) was used with the standard E-value cutoff of 1. For each assayed putative insect allergen, the number of positive hits for both the global (80mer window) and the local (8mer window) identities is indicated in **Table 1**.

Insect	Putative allergen	(No. hits 80mer)	(No. hits 8mer)
Tenebrio molitor	Alpha-amylase	6	41
	Chitinase	2	0
	Cockroach allergen	10	0
	Glutathione S-transferase	3	57
	HSP 70	7	389
	Hexamerin	9	37
	Serine protease	11	0
	Triosephosphate isomerase	4	132
Bombyx mori	Actin	0	0
	Alpha-amylase	4	43
	Arginine kinase	14	1431
	Chitinase	2	3
	Glutathione S-transferase	4	24
	HSP 70	7	218
	Hemocyanin	9	12
	Sarcoplasmic Ca-binding protein	4	6
	Serine protease	3	0
	Triosephosphate isomerase	4	107
	Tropomyosin	75	866
	Troponin C	12	115
	Trypsin	10	0
	Beta-tubulin	0	0
Musca domestica	Actin	0	0
	Alpha-amylase	6	25
	Arginine kinase	14	1045
	Chitinase	2	0
	Glutathione S-transferase	3	9
	HSP 70	7	633
	Hemocyanin	9	3
	Sarcoplasmic Ca-binding protein	0	0
	Serine protease	14	6
	Triosephosphate isomerase	4	106
	Tropomyosin	76	4547
	Troponin C	10	19
	Trypsin	15	19
	Beta-tubulin	0	0

Insect	Putative allergen	(No. hits 80mer)	(No. hits 8mer)
Locusta migratoria	Actin	0	0
	Arginine kinase	14	1329
	Chitinase	2	0
	Glutathione S-transferase	3	4
	HSP 70	8	602
	Hexamerin	9	0
	Serine protease	16	13
	Tropomyosin	76	5455
	Trypsin	16	13
	Beta-tubulin	0	0
Acheta domesticus	Triosephosphate isomerase	4	27
Galleria mellonella	Glutathione S-transferase	3	22
	Hemocyanin	9	23
	Trypsin	16	12
Hermetia illucens	Alpha-amylase	6	56
	Serine protease	16	19
	Trypsin	16	57
Schisto americana	Arginine kinase	14	1383

Table 1. Global (80mer) and local (8mer) identities found for the putative insect allergens.

3. IgE-binding allergens of edible insects belong to conserved protein families

Bioinformatic investigations using a sliding window of 80 amino acids resulted in a large number of positive hits for the putative allergen proteins of all the insect species, with the exception of actin, sarcoplasmic Ca-binding protein (SCBP), and β-tubulin (**Table 1**). However, some of the global identities do not necessarily coincide with local identities, since no hit was found with a more restricted sliding window of eight amino acid residues. Both global and local identities were found with the thioredoxin allergen of silkworm, house fly maggot, and the Indian mealmoth (*P. interpunctella*).

Most of the hits found with the 80mer and 8mer windows consist of allergens from arthropods such as dust mites, crustaceans, and insects and, more rarely, of allergens from mollusks, nematodes, and fungi (*Alternaria alternata, Aspergillus oryzae, Aspergillus fumigatus, Cladosporium herbaceum, Malassezia sympodialis*) (**Table 2**). For a limited number of putative insect allergens like the widely distributed heat shock protein HSP 70, serine protease, trypsin, triosephosphate isomerase (TPI), and thioredoxin, hit allergens of plant (the blue cypress *Cupressus arizonica*, the maize *Zea mays*, the wheat *Triticum aestivum*, the common ragweed *Ambrosia artemisiifolia*, the olive tree *Olea europaea*) and animal (the Mozambique tilapia *Oreochromis mossambicus*, the dog *Canis familiaris*) origin were identified. Obviously, these allergens consist of ubiquitous pan-allergens that occur in so distantly phylogenetically related or phylogenetically unrelated organisms.

Protein family	Insects	Dust mites	Crustaceans	Mollusks/Nematodes
Actin	–	(1)	–	–
Alpha-amylase	Bla g 11	Blo t 4	–	–
	Per a 11	Der f 4		
	(+2)	Der p 4		
		Eur m 4 (+2)		
Arginine kinase	Bomb m 1	Der f 20	Cra c 2	(4)
	Per a 9	Der p 20 (+3)	Lit v 2	
	Plo i 1 (+5)		Pen m 2 (+23)	
Chitinase	Per a 12 (+1)	Der f 15 (+5)	–	–
Glutathione	Bla g 5 (+1)	Blo t 8	–	–
S-transferase		Der f 8		Asc l 13 (N)
		Der p 8 (+6)		Asc s 13 (N)
HSP 70 (heat shock protein)	Aed a 8 (+2)	Der f 28	–	–
		Tyr p 28		
Hemocyanin	Bla g 3	–	(1)	–
	Per a 3			
Hexamerin	(6)	–	–	–
Myosin	Bla g 8 (+1)	Der f 26	Art fr 5	–
			Cra c 5	–
			Hom a 3	
			Lit v 3	
			Pen m 3 (+1)	
Sarcoplasmic Ca-binding protein	Aed a 5 (+2)	–	Cra c 4	–
			Eri s 4	
			Lit v 4	
			Mac r 4	
			Pen m 4	
			Pon l 4 (+24)	
Serine protease	Api m 7	Der f 6	–	–
	Bom t 4	Der p 6		
	Per a 10	Eur m 1 (+12)		

Protein family	Insects	Dust mites	Crustaceans	Mollusks/Nematodes
Triosephosphate isomerase	Pol d 4			
	Pol e 4 (+14)			
Tropomyosin	(2)	Der f 25	Arc s 8, Cra c 8	–
	Aed a 10	Blo t 10	Cha f 1	Ani s 3 (N)
	Bla g 7	Cho a 10	Cra c 1	Asc l 3 (N)
	Chi k 10	Der f 10	Hom a 1	Hel as 1 (+50)
	Lep s 1	Der p 10	Lit v 1	
	Per a 7 (+29)	Lep d 10	Mac r 1	
		Tyr p 10 (+11)	Mel l 1	
			Met e 1	
			Pan s 1	
			Pen a 1	
			Pen m 1	
			Por p 1 (+54)	
Troponin C	Bla g 6, Per a 6	Tyr p 34	Cra c 6	
			Hom a 6	(1 N)
			Pen m 6 (+2)	
Trypsin	(4)	Blo t 3	–	–
		Der f 3		
		Der p 3		
		Eur m 3		
		Tyr p 3 (+3)		
Alpha-tubulin	–	Der f 33 (+2)	–	–

The Internaional Union of Immunological Societies (IUIS) nomenclature (the three first initial of the genus name, followed by the initial of the species name, followed by a number indicating the ranking of discover, e.g., **Lit v 2** for the shrimp ***Lit**openaeus **v**annamei* **2** allergen) was used. Allergens referenced by IUIS (2016) are indicated; numbers into brackets represent other allergens nonreferenced by IUIS but included into the AllergenOnline data bank.

Table 2. Nomenclature of the IgE-binding allergens belonging to the main allergenic protein families identified in insects, dust mites, crustaceans, mollusks, and nematodes (N) .

All the insect allergens identified so far consist of proteins which belong to families of highly conserved proteins, namely, muscle proteins such as tropomyosin, myosin, and the sarcoplasmic Ca-binding protein and enzymes such as α-amylase, chitinase, glutathione S-transferase (GST), arginine kinase, serine protease, and trypsin. Most of these proteins have been already identified as allergens of both the German (*Blattella germanica*) and American cockroach (*Periplaneta americana*) (http://Allergome.org). According to the high degree of conservation, all of these allergens are distributed among phylogenetically related clusters in

http://dx.doi.org/10.5772/68124

the phylogenetic trees built up from their amino acid sequence alignments. As an example, **Figure 1** illustrates the phylogenetic tree built up from the glutathione S-transferase multiple alignment. Very similar trees were built up from the multiple alignments of other enzyme allergens from edible insects, crustaceans, mollusks, and nematodes (results not shown), except for the tropomyosin tree, in which insect tropomyosins cluster in two separate groups

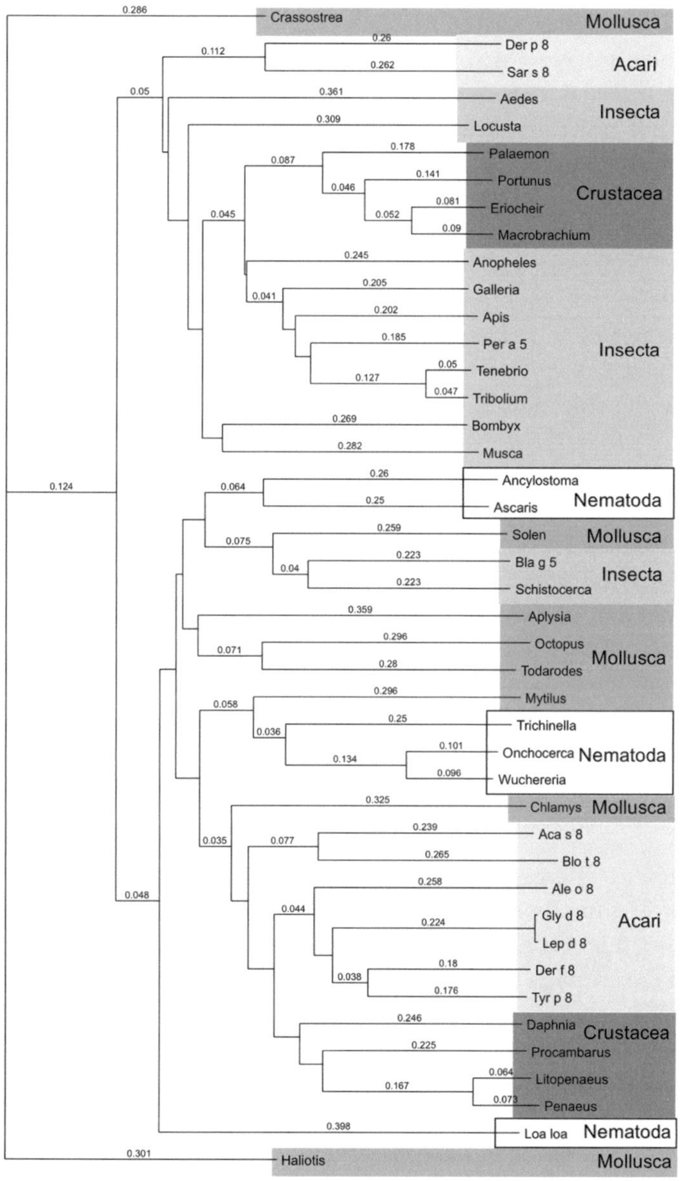

Figure 1. Phylogenetic tree built up from the amino acid sequence alignment of glutathione S-transferase allergens of dust mites, insects, crustaceans, mollusks, and nematodes. Clusters corresponding to dust mites, insects, crustaceans, mollusks, and nematodes are differently shaded.

that are differently phylogenetically related to the dust mite tropomyosin cluster [11]. This discrepancy observed in the distribution of insect tropomyosins is consistent with the fact that some of the dust mite tropomyosin-reactive patient sera strongly interacted with a tropomyosin-containing mealworm extract in western blot experiments, whereas other dust mite tropomyosin-reactive patient sera did not react at all [11].

3.1. Muscle proteins

The muscle proteins tropomyosin, myosin, and sarcoplasmic Ca-binding protein (SCBP) have been identified as major allergens of edible insects. Especially, tropomyosin appears as a major pan-allergen largely distributed among dust mites, insects, crustaceans, mollusks, and nematodes [12–16]. Major allergens of dust mites, e.g., Aca s 10 from *Acarus siro*, Blo t 10 from *Blomia tropicalis*, Der f 10 and Der p 10 from *Dermatophagoides farinae* and *Dermatophagoides pteronyssinus*, Gly d 10 from *Glycyphagus destructor*, Ixo sc 10 from the shoulder tick *Ixodes scapularis*, and Tyr p 10 from *Tyrophagus putrescentiae*, consist of tropomyosins (Allergome.org). Many other tropomyosins consist of the major allergens of insects, e.g., Bla g 7 and Per a 7 from the cockroaches *B. germanica* and *P. americana*, Bomb m 7 from the edible pupa of *B. mori*, Aed a 7, and Cul q 7 from the mosquitos *Aedes aegypti* and *Culex quinquefasciatus*, and Chi k 10 from the chironomid *Chironomus kiiensis*, Dro m 7 from the fruit fly *Drosophila melanogaster*, Glo m 7 from the tsetse fly *Glossina morsitans*, and Loc m 7 from the edible locust *L. migratoria* (Allergome.org). Many crustacean species also contain tropomyosin as a major muscle allergen, e.g., Cra c 1 from the common shrimp *Crangon crangon*, Eri s 1 from the crab *Eriocheir sinensis*, Hom a 1 from the American lobster *Homarus americanus*, Lit v 1 and Pen m 1 from the prawns *Litopenaeus vannamei* and *Penaeus monodon*, Nep n 1 from the scampi *Nephrops norvegicus*, etc. (Allergome.org). Tropomyosin also occurs as a major allergen in various mollusks like Cra g 1 from the oyster *Crassostrea gigas*, Hel a from the snail *Helix aspersa*, Hal a 1 from the abalone *Haliotis asinina*, Lol b 1 from the spear squid *Loligo bleekeri*, Myt e 1 from the blue mussel *Mytilus edulis*, Oct v 1 from *Octopus vulgaris*, Por t 1 from the Japanese blue crab *Portunus trituberculatus*, and Sep of 1 from the common cuttlefish *Sepia officinalis* (Allergome.org). Finally, tropomyosin also consists of the major allergen Ani s 3 of the nematode *Anisakis simplex* (http://Allergome.org).

Other muscle protein allergens like troponin and the sarcoplasmic Ca-binding protein (SCBP) also provide a number of allergens like the troponins Tyr p 24 from the dust mite *T. putrescentiae*, Bla g 6 and Per a 6 from the cockroaches *B. germanica* and *P. americana*, Cra c 6 and Pen m 6 from the shrimps *C. crangon* and *P. monodon*, Hom a 6 from the American lobster *H. americanus*, and the troponin of the nematode *A. simplex* (Allergome.org). The sarcoplasmic Ca-binding protein also occurs as an allergen in insects (Aed a 4 and Cul q 4 from the mosquitos *Aedes aegypti* and *C. quinquefasciatus*) and crustaceans (Cra c 4 from *C. crangon*, Eri s 4 from the Chinese crab *E. sinensis*, Hom a 4 from the lobster *H. americanus*, Mac r 4 from the giant freshwater prawn *Macrobrachium rosenbergii*, Pen m 4 from *P. monodon*, Scy pa 4 from the green mud crab *Scylla paramamosain*) (Allergome.org). To date, no SCBP has been identified as an allergen in mollusks and nematodes.

All of these muscle protein allergens display a rather high resistance to both the proteolysis and heat denaturation, as exemplified by the experiments performed on the tropomyosin of different species of mealworm [17] and the oyster *Crassostrea gigas* [18].

3.2. Enzymes

A number of enzymes including hydrolases like α-amylase, chitinase, serine protease, and trypsin and metabolic enzymes like arginine kinase (AK), glutathione S-transferase (GST), and triosephosphate isomerase (TPI) have been identified as cross-reacting allergens of edible insects [11, 19–23].

Arginine kinase has been previously identified as a pan-allergen widely distributed in various insects such as the yellow mealworm (*T. molitor*) [20], the field cricket (*Gryllus bimaculatus*) [23], and the house cricket (*A. domesticus*) [11]; shrimps like the black tiger prawn (*P. monodon*) and the king prawn (*Penaeus latisulcatus*) [24]; the giant freshwater prawn (*M. rosenbergii*) [23, 25]; and crabs like the blue swimming crab (*Portunus pelagicus*) [26] and the red crab (*c*) [27]. Arginine kinases consist of the major allergens Bla g 9 of the German cockroach (*B. germanica*) and Per a 9 of the American cockroach (*P. americana*) (Allergome.org). Many other allergens of dust mites like Blo t 20 of *B. tropicalis*, Der f 20 of *D. farinae*, Der p 20 of *D. pteronyssinus*, and Gly d 20 of *G. destructor* also consist of arginine kinases (Allergome.org). The list of arginine kinase allergens of crustaceans is also consistent (http://Allergome.org).

Alpha-amylase, a hydrolase of paramount importance for the digestion of starch by herbivorous and omnivorous organisms, occurs as an allergens in dust mites (Aca s 4 of *A. siro*, Blo t 4 of *B. tropicalis*, Der p 4 of *D. pteronyssinus*, Eur m 4 of *Euroglyphus maynei*, Tyr p 4 of *T. putrescentiae*) and insects (Sim v 3 and Sim v 4 of the striped black fly *Simulia vittata*, Bla g 11 and Per a 11 of the cockroaches *B. germanica* and *P. americana*) (http://Allergome.org).

Other metabolic enzymes like the glutathione S-transferase GST (Aca s 8 of *A. siro*, Blo t 8 of *B. tropicalis*, Der f 8 and Der p 8 of *D. farinae* and *D. pteronyssinus*, Tyr p 8 of *Tyroglyphus putrescentiae*, Bla g 5 and Per a 5 of *B. germanica* and *P. americana*), chitinase (Blo t 15 of *B. tropicalis*, Der f 15 and Der p 15 of *D. farinae* and *D. pteronyssinus*, and Per a 12 of the cockroach *P. americana*), and triosephosphate isomerase TPI (Der f 25 of *D. farinae*, Bla g TPI of the cockroach *B. germanica*, For t TPI of the biting midge *Forcipomyia taiwana*, and Cra c 8 of the shrimp *C. crangon*) also consist of allergens essentially in dust mites and insects (Allergome.org). However, they seem to be less widely distributed in arthropods than other enzymes like arginine kinase.

3.3. Other proteins

Other proteins involved in metabolic pathways (HSP70, thioredoxin) or displaying structural (tubulin) or physiological (hemocyanin and hexamerin) functions also occur as minor allergens in edible insects (**Table 2**). The hemolymph proteins hemocyanin and hexamerin both consist of homotetrameric proteins of high molecular mass, which share very conserved amino acid sequences and three-dimensional conformations. Hexamerin is widely distributed

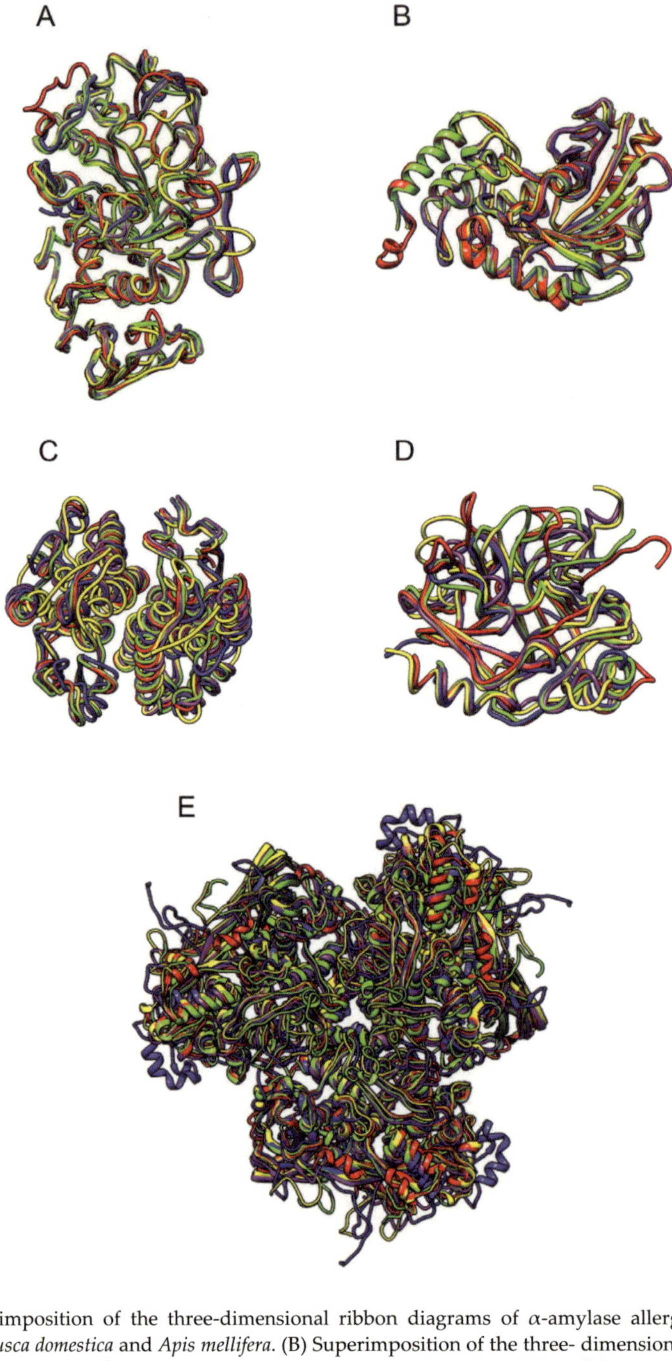

Figure 2. (A) Superimposition of the three-dimensional ribbon diagrams of α-amylase allergens of *Bombyx mori, Hermetia illucens, Musca domestica* and *Apis mellifera*. (B) Superimposition of the three- dimensional ribbon diagrams of arginine kinase allergens of *Bombyx mori, Locusta migratoria, Musca domestica, Apis mellifera* and *Schistocerca americana*. (C) Superimposition of the three-dimensional ribbon diagrams of glutathione S-transferase allergens of *Tenebrio molitor, Locusta migratoria, Galleria mellonella, Musca domestica* and *Apis mellifera*. (D) Superimposition of the three-dimensional ribbon diagrams of trypsin allergens of *Bombyx mori, Galleria mellonella, Hermetia illucens, Locusta migratoria* and *Musca domestica*. (E) Superimposition of the three-dimensional ribbon diagrams of hexamerin allergens of *Bombyx mori, Locusta migratoria, Galleria mellonella, Tenebrio molitor* and *Schistocerca americana*.

among insects and crustaceans, and it has been identified as an allergen of the fly maggot [28]. The hemolymph protein hemocyanin has been identified as an allergen of the German cockroach (Bla g 3) and American cockroach (Per a 3) and of the giant freshwater prawn *M. rosenbergii* as well (Allergome.org).

Owing to the conserved character, all the IgE-binding cross-reacting allergens of edible insects share very similar and readily superposable three-dimensional conformations. These structural similarities are illustrated in **Figure 2**, which shows the nice superposition of α-amylase, arginine kinase, glutathione S-transferase, trypsin, and hexamerin models of different origins. In fact, as shown for most of the members in different groups of evolutionary-related proteins, the three-dimensional conformations are much more conserved than the corresponding amino acid sequences.

4. Resistance of the insect allergens to proteolysis by digestive enzymes

Resistance to proteolysis consists of a property of paramount importance for food allergens, allowing them to escape the proteolytic degradation along the digestive tract and, thus, preserving their ability to stimulate the peripheral lymph nodes, e.g., Peyer's patches, associated with the intestinal tract. In this respect, all of the putative insect allergenic enzymes such as α-amylase, arginine kinase, glutathione S-transferase, and trypsin exhibit a number of predicted cleavage sites by pepsin and trypsin distributed along their polypeptide chain and, especially, exposed on their molecular surface (**Figure 3**). Accordingly, the multiple proteolysis of all of these enzymes by pepsin, trypsin, and chymotrypsin generate a number of amino acids and short peptides apparently devoid of efficient IgE-binding properties (**Figure 3**). However, a limited number of peptides would keep a sufficient size (>10 amino acid residues), to properly stimulate the digestive immune system. In this respect, the allergenicity of tropomyosin, myosin, α-amylase, and hexamerin from the yellow mealworm (*T. molitor*), the giant mealworm beetle (*Zophobas atratus*), and the litter beetle (*Alphitobius diaperinus*) was reduced but not abolished following both in vitro simulated gastric fluid (SGF) and in vitro simulated intestinal fluid (SIF) digestion and heat treatment [17, 29]. The heat resistance of the major allergens of edible insects implies that both cooked insects and insect protein-containing food products retain some intact allergenicity. Heat and proteolysis stability of tropomyosin from the mud crab (*Scylla serrata*) [30] and the tropical oyster *Crassostrea belcheri* [18] have been similarly pointed out.

5. What extent for the allergy to edible insects?

To date, only a few cases of allergenic manifestations caused by the consumption of edible insects have been reported in the literature. The first case reports deal with occupational allergies of particularly exposed environmental searchers, fishers, and food industry workers [21, 28, 31–37]. Similarly, the well-known "pancake syndrome" (oral mite anaphylaxis), caused by the unintended consumption of mite-contaminated foods, has been identified in Refs. [38, 39]. Interestingly, most or less severe cases of anaphylaxis caused by the ingestion of various edible insects, reported in Chinese journals [40–51], were collated by Ji et al. [52], who counted up to 358 episodes of anaphylactic shock caused by food ingestion from 1980 to 2007. The most common offending allergens were identified as pineapple (25%), the soft-shelled turtle (Trionychidae) (19%), crabs (9%), and edible insects (locust + grasshopper)

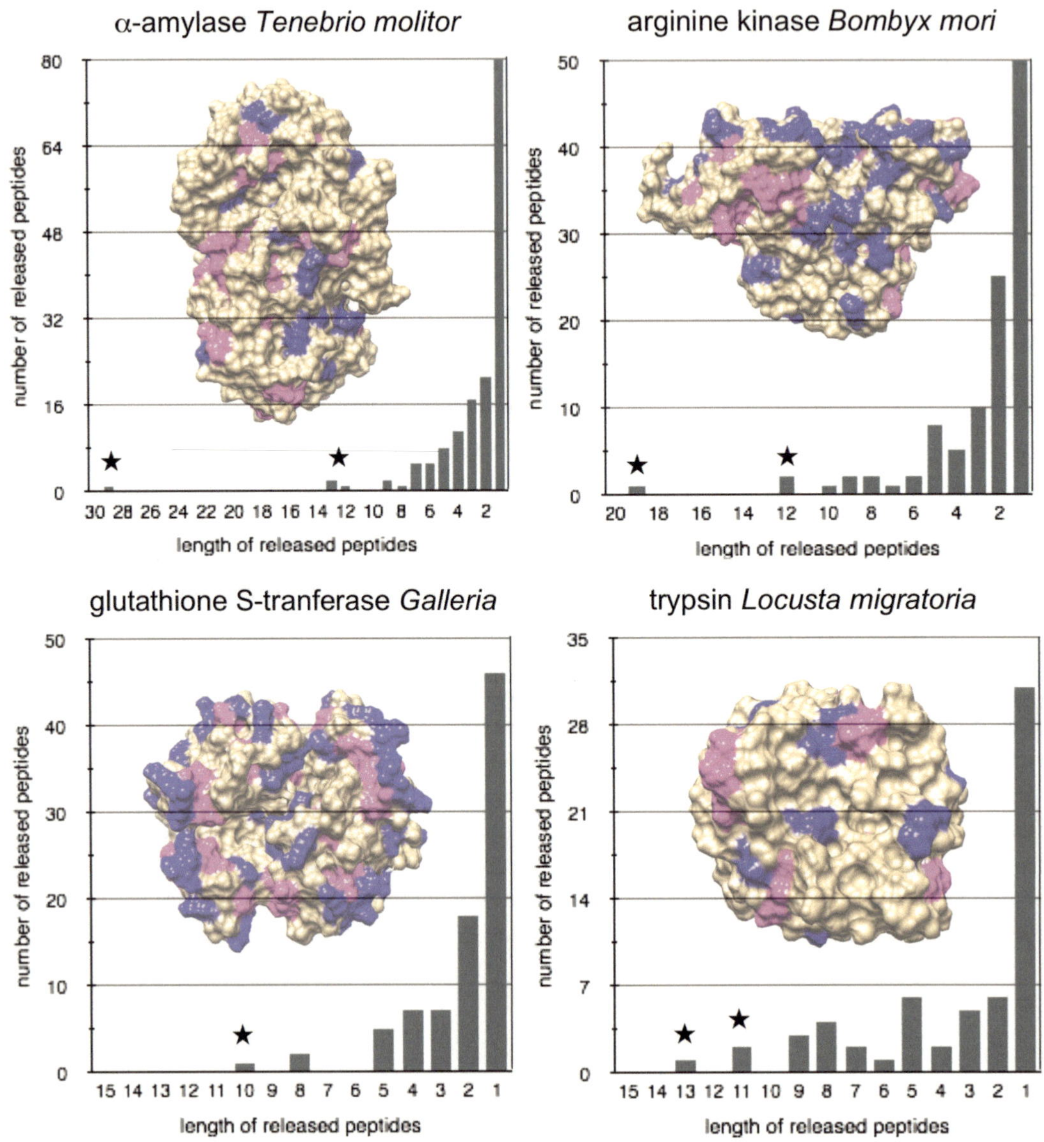

Figure 3. Size (number of amino acid residues) diagram of the peptides resulting from the predicted multiple proteolysis by pepsin, trypsin, and chymotrypsin of α-amylase of *Tenebrio molitor*, arginine kinase of *Bombyx mori*, glutathione S-transferase of *Galleria mellonella*, and trypsin from *Locusta migratoria*. Peptides of ≥10 amino acid residues in length are indicated by stars (★). Overlay images showing the localization of the predicted cleavage sites by pepsin (pale grey) and trypsin (deep grey) on the molecular surface of the corresponding allergens are presented.

(14%). Other cases of anaphylaxis caused by the ingestion of edible insects were subsequently reported, mainly in Asia [53–55]. More recently, a majority of shrimp-allergic people (13 over 15) were confirmed as being allergic to yellow mealworm (*T. molitor*) when tested in double-blind, placebo-controlled food challenge (DBPCFC) [56]. In this respect, yellow mealworm appears as a food at least as allergenic as shrimps to trigger anaphylactic responses in shrimp-allergic patients.

The limited number of reported cases of anaphylaxis due to edible insect consumption seems to be largely underestimated, especially in countries like China, where a great variety of insects are traditionally consumed as a source of dietary proteins. The occurrence in all of the edible insects of IgE-binding allergens which cross-react with the major allergens tropomyosin and arginine kinase of shrimps, dust mites, mollusks, and even nematodes suggests that shrimp-allergic and mollusk-allergic patients are at risk when consuming edible insects or insect-containing food products. However, further large-scale investigations among a broad population of shrimp- and mollusk-allergic patients will be necessary to appreciate the real allergenic risk edible insects pose to previously sensitized individuals. In the meantime, it would be wise to inform the consumers for such a potential risk, e.g., by a proper labeling of insect foods and insect-containing food products.

6. Conclusion

Obviously, the repertoire of food allergens from edible insects consists of a number of IgE-binding cross-reactive allergens common to other arthropods, e.g., dust mites and crustaceans, mollusks, and, more scarcely, nematodes. These pan-allergens refer to muscle proteins, enzymes, and proteins with structural and physiological functions. However, the search of identities the insect proteins share with known allergens of the allergen bank as a criterion for identifying allergens of edible insects suffers from some limitations associated to the completeness and quality of the bank. Most of the allergenic proteins of animal or plant origin essentially belong to abundant and widespread protein families in both animal and plant species like tropomyosins, lipocalins, and caseins for animals and cupins, profilins, and prolamins for plants [57]. Moreover, depending on the data bank used for searching the identities with known allergens, the accuracy and exhaustiveness of the results might vary considerably. In this respect, the continuously updated FARRP AllergenOnline bank offers a maximum of guarantee for the retrieved information [58]. Accordingly, all of the allergens identified to date correspond to proteins already known for their allergenic properties. Other allergens more specific of insects remain to be identified and characterized, in order to have a more accurate idea about the variability and specificity of the edible insect allergens. A serological approach using IgE-containing sera from allergic patients will be necessary to fulfill such a requirement, instead of the computational approach reported in this chapter. As insect food could be so allergenic that it can trigger strong anaphylactic responses in allergic persons, it is recommended that all insect food and insect-containing food products should mention this allergy possibility very clearly in the product labels.

Author details

Pierre Rougé* and Annick Barre

*Address all correspondence to: pierre.rouge@free.fr

University Paul Sabatier – Tolouse 3, Research Institute for Development, Research Unit 152 Pharma-Dev, Faculty of Pharmacy, Toulouse, France

References

[1] Rumpold BA, Schluter OK. Potential and challenges of insects as an innovative source for food and feed production. Innovative Food Science & Emerging Technologies. 2013;**17**:1-11.

[2] DeFoliart G, Nakagaki B, Sunde M. Protein quality of the house cricket *Acheta domestica* when fed to broiler chicks. Poultry Science. 1987;**66**:1367-1371.

[3] Belluco S, Losasso C, Maggioletti M, Alonzi CC, Paoletti MG, Ricci A. Edible insects in a food safety and nutritional perspective: a critical review. Comprehensive Reviews in Food Science and Food Safety. 2013;**12**:296-313.

[4] Rumpold BA, Schluter OK. Nutritional composition and safety aspects of edible insects. Molecular Nutrition & Food Research. 2013;**57**:802-823.

[5] Li L, Zhao Z, Liu H. Feasibility of feeding yellow mealworm (*Tenebrio molitor* L.) in bioregenerative life support systems as a source of animal protein for humans. Acta Astronautica. 2013;**92**:103-109.

[6] van Huis A, Van Itterbeeck J, Klunder H, Mertens E, Halloran A, Muir G, Vantomme P. Edible Insects: Future Prospects for Food and Feed Security. FAO Forestry Paper 171, ISBN 978-92-5-107595-1. Rome: Food and Agriculture Organization of the United Nations; 2013.

[7] Charlton AJ, Dickinson M, Wakefield, ME, Fitches E, Kenis M, Han R, Zhu F, Kone N, Grant M, Devic E, Bruggeman G, Prior R, Smith R. Exploring the chemical safety of fly larvae as a source of protein for animal feed. Journal of Insects as Food and Feed. 2015;**1**: 7-16.

[8] Van der Brempt X, Moneret-Vautrin DA. The allergic risk of *Tenebrio molitor* for human consumption (article in French). Revue Française d'Allergologie. 2014;**54**:34-36.

[9] Barre A, Caze-Subra S, Gironde C, Bienvenu F, Bienvenu J, Rougé P. Entomophagy and the risk of allergy (article in French). Revue Française d'Allergologie. 2014;**54**:315-321.

[10] Van der Brempt X, Beaudoin E, Lavaud F. Is eating insects risky for allergic patients? (article in French). Revue Française d'Allergologie. 2016;**56**:186-188.

[11] Barre A, Velazquer E, Delplanque A, Caze-Subra S, Bienvenu F, Bienvenu J, Benoist H, Rougé P. Cross-reacting allergens of edible insects (article in French). Revue Française d'Allergologie. 2016;**56**:522-532.

[12] Daul CB. Slattery M, Reese G, Lehrer SB. Identification of the major brown shrimp (*Penaeus aztecus*) allergen as the muscle protein tropomyosin. International Archives of Allergy and Immunology. 1994;**105**: 49-55.

[13] Santos ABR, Chapman MD, Aalberse RC, Vailes LD, Ferriani VPL, Oliver C, Rizzo MC, Naspitz CK, Arruda LK. Cockroach allergens and asthma in Brazil: Identification of tropomyosin as a major allergen with potential cross-reactivity with mite and shrimp allergens. The Journal of Allergy and Clinical Immunology. 1999;**104**:329-337.

[14] Reese G, Ayuso R, Lehrer SB. Tropomyosin: An invertebrate pan-allergen. International Archives of Allergy and Immunology. 1999;**119**:247-258.

[15] Galindo PA, Lombardero M, Borja JEG, Feo F, Barber D, García R. A new arthropod panallergen? Allergy. 2001;**56**:195-197.

[16] Shafique RH, Inam M, Ismail M, Chaudhary FR. Group 10 allergens (tropomyosins) from house-dust mites may cause covariation of sensitization to allergens from other invertebrates. Annergy Rhinology. 2010:**3**:e74-e90.

[17] Broekman H, Knulst A, den Hartog Jager S, Monteleone F, Gaspari M, de Jong G, Houben G, Verhoeckx K. Effect of thermal processing on mealworm allergenicity. Molecular Nutrition & Food Research. 2015;**59**:1855-1864.

[18] Yadzir ZH, Misnan R, Bakhtiar F, Abdullah N, Murad S. Tropomyosin, the major tropical oyster *Crassostrea belcheri* allergen and effect of cooking on its allergenicity. Allergy, Asthma and Clinical Immunology. 2015;**11**:30.

[19] Liu Z, Xia L, Wu Y, Xia Q, Chen J, Roux KH. Identification and characterization of an arginine kinase as the major allergen from silkworm (*Bombyx mori*) larvae. International Archives of Allergy and Immunology. 2009;**150**:8-14.

[20] Verhoeckx KCM, van Broekhoven S, den Hartog-Jager CF, Gaspari M, de Jong GAH, Wichers HJ, van Hoffen E, Houben GF, Knulst AC. House dust mite (Der p 10) and crustacean allergic patients may react to food containing yellow mealworm proteins. Food and Chemical Toxicology. 2014;**65**:364-373.

[21] Debaugnies F, Francis F, Delporte C, Doyen V, Lendent C, Mairesse M, Van Antwerpen P, Corraza F. Identification de l'α-amylase comme allergène du ver de farine chez des patients professionnellement exposés (article in French). Revue Française d'Allergologie. 2016;**56**:281.

[22] Binder M, Mahler V, Hayek B, Sperr WR, Schöller M, Prozell S, Wiedermann G, Valent P, Valenta R, Duchêne M. Molecular and immunological characterization of arginine kinase from the Indian meal moth, *Plodia interpunctella*, a novel cross-reactive invertebrate pan-allergen. Journal of Immunology. 2001;**167**:5470-5477.

[23] Srinroch C, Srisomsap C, Chokchaichamnankit D, Punyarit P, Phiriyangkul P. Identification of novel allergen in edible insect, *Gryllus bimaculatus* and its cross-reactivity with *Macrobrachium* spp. allergens. Food Chemistry. 2015;**184**:160-166.

[24] Sahabudin S, Misnan R, Yadzir ZH, Mohamad J, Abdullah N, Bakhtiar F, Murad S. Identification of major and minor allergens of black tiger prawn (*Penaeus monodon*) and king prawn (*Penaeus latisulcatus*). The Malaysian Journal of Medical Sciences. 2011;**18**:2732.

[25] Yadzir ZHM, Misnan R, Abdullah N, Bakhtiar F, Arip M, Murad S. Identification of the major allergen of *Macrobrachium rosenbergii* (giant freshwater prawn). Asian Pacific Journal of Tropical Biomedicine. 2012;**2**:50-54.

[26] Rosmilah M, Shahnaz M, Zailatul HMY, Noormalin A, Normilah I. Identification of tropomyosin and arginine kinase as major allergens of *Portunus pelagicus* (blue swimming crab). Tropical Biomedicine. 2012;**29**:467-478.

[27] Misnan R, Murad S, Yadzir ZH, Abdullah N. Identification of the major allergens of *Charybdis feriatus* (red crab) and its cross-reactivity with *Portunus pelagicus* (blue crab). Asian Pacific Journal of Allergy and Immunology. 2012;**30**:285-293.

[28] Just N, Lièvre K, Lallemand K, Leduc V. Implication de l'hexamérine dans un cas d'allergie aux larves de mouches (article in French). Revue Française d'Allergologie. 2012;**52**:258.

[29] Van Broekhoven S, Bastiaan-Net S, de Jong NW, Wichers HJ. Influence of processing and *in vitro* digestion on the allergic cross-reactivity of three mealworm species. Food Chemistry. 2016;**196**:1075-1083.

[30] Huang YY, Liu GM, Cai QF, Weng WY, Maleki SJ, Su WJ, Cao MJ. Stability of major allergen tropomyosin and other food proteins of mud crab (*Scylla serrata*) by *in vitro* gastrointestinal digestion. Food and Chemical Toxicology. 2010;**48**:1196-1201.

[31] Bernton HS, Brown H. Insects as potential sources of ingestant allergens. Annals of Allergy. 1967;**25**:381-387.

[32] Meier-Davis S, Bush RK. Occupational sensitivity to *Tenebrio molitor* Linnaeus (yellow mealworm). The Journal of Allergy and Clinical Immunology. 1990;**86**:182-188.

[33] Phillips J, Burkholder W. Allergies related to food insect production and consumption. The Food Insects Newsletter. 1995;**8**:1-2.

[34] Teranishi H, Kawai K, Murakami G, Miyao M, Kasuya M. Occupational allergy to adult chironomid midges among environmental researchers. International Archives of Allergy and Immunology. 1995;**106**:271-277.

[35] Freye HB, Esch RE, Litwin CM, Sorkin L. Anaphylaxis to the ingestion and inhalation of *Tenebrio molitor* (mealworm) and *Zophobas morio* (superworm). Allergy and Asthma Proceedings. 1996;**17**:215-219.

[36] Aldunate MT, Echechipía S, Gómez B, García BE, Olaguibel JM, Rodríguez A, Moneo I, Tabat AI. Chironomids and other causes of fish food allergy. Alergología e inmunología clínica. 1999;**14**:140-145.

[37] Meseguer Arce J, Sánchez-Guerrero Villajos IM, Iraola V, Carnés J, Fernández Caldas E. Occupational allergy to aquarium fish food: Red midge larva, freshwater shrimp, and earthworm. A clinical and immunological study. Journal of Investigational Allergology and Clinical Immunology. 2013;**23**:462-470.

[38] Sánchez-Borges M, Capriles-Hulett A, Fernández-Caldas E, Suárez-Chacón R, Caballero F, Castillo S, Sotillo E. Mite-contaminated foods as a cause of anaphylaxis. The Journal of Allergy and Clinical Immunology. 1997;**99**:738-743.

[39] Sánchez-Borges M, Suárez-Chacón R, Capriles-Hulett A, Caballero-Fonseca F, Iraola V, Fernández-Caldas E. Anaphylaxis from ingestion of mites: Pancake anaphylaxis. The Journal of Allergy and Clinical Immunology. 2013;**131**:31-35.

[40] Wei SZ, Jia LL, Wang SX. A case of anaphylactic shock caused by silkworm pupa (article in Chinese). Chinese Journal of Medicine. 1981;**299**:43.

[41] Cheng XL, Lin HZ, Zhang SY. Successful curative treatment of a patient with anaphylactic shock caused by ingestion of silk worm pupa (article in Chinese). Guangzhou Med. 1987;**3:**36.

[42] Wang YL, Niu CX, Liu Y. Three cases of anaphylactic shock caused by ingestion of silkworm pupa (article in Chinese). China Journal of Leprosy and Skin Diseases. 1999;**15**:56-57.

[43] Qiao X. A case of anaphylactic shock caused by locust consumption (article in Chinese). Medical Journal of the Chinese People's Armed Police Forces. 1999;**7**:414.

[44] Zhang H. A case of anaphylactic shock caused by ingestion of silkworm pupa (article in Chinese). Lit. Inf. Prev. Med. 2002;**8**:458-459.

[45] Pao DH, Zhao, GS. A case of anaphylactic shock caused by cicada pupa consumption (article in Chinese). People's Military Surgeon. 2003;**46**;246.

[46] Wan YF, Wei ZF. A case of anaphylactic shock caused by bee pupa consumption (article in Chinese). Clinical Journal of Medical Officer. 2003;**31**:52.

[47] Wang Y, Feng SH. A case of anaphylactic shock caused by ingestion of bee larva (article in Chinese). Medical Journal of National Defending Forces in Southern China. 2003;**13**:346.

[48] Wang DJ, Zhang DL, Yang FX, Chen YZ, Wei L, Yang HX et al. Eight cases of severe type-1 allergic reaction caused by consumption of silkworm pupa (article in Chinese). Henan Journal of Preventive Medicine. 2005;**16**:148.

[49] Liu XP. Nineteen cases of anaphylactic shock caused by grasshopper consumption (article in Chinese). China Modern Doctor. 2007;**45**:41.

[50] Zhang YZ. A case of anaphylactic reaction caused by ingestion of *Clanis bilineata* (article in Chinese). Clinical Wrong Diagnosis and Wrong Therapy. 2007;**20**:125.

[51] Ji KM, Zhan ZK, Chen JJ, Liu ZG. Anaphylactic shock caused by silkworm pupa consumption in China (article in Chinese). Allergy. 2008;**63**:1407-1408.

[52] Ji K, Chen J, Li M, Liu Z, Wang C, Zhan Z, Wu X, Xia Q. Anaphylactic shock and lethal anaphylaxis caused by food consumption in China. Trends in Food Science and Technology. 2009;**20**:227-231.

[53] Choi GS, Shin Y, Kim JE, Ye YM, Park HS. Five cases of food allergy to vegetable worm (*Cordyceps sinensis*) showing cross-reactivity with silkworm pupae. Allergy. 2010;**65**:1196-1197.

[54] Yew KL, Kok VSL. Exotic food anaphylaxis and the broken heart: sago worm and Takotsubo cardiomyopathy. The Medical Journal of Malaysia. 2012;**5**:540-541.

[55] Okezie OA, Kgomotso KK, Letswiti MM. Mopane worm allergy in a 36-year-old woman: a case report. Journal of Medical Case Reports. 2010;**4**:42.

[56] Broekman H, Verhoeckx KC, den Hartog Jager CF, Kruizinga AG, Pronk-Kleinjan M, Remington BC, Bruijnzeel-Koomen CA, Houben GF, Knulst AC. Majority of shrimp-allergic patients are allergic to mealworm. The Journal of Allergy and Clinical Immunology. 2016;**137**:1261-1263.

[57] Radauer C, Bublin M, Wagner S, Mari A, Breiteneder H. Allergens are distributed into few protein families and possess a restricted number of biochemical functions. The Journal of Allergy and Clinical Immunology. 2008;**121**:847-852.

[58] Sircar G, Sarkar D, Bhattacharya SG, Saha S. Allergen databases. Methods in Molecular Biology. 2014;**1184**:165-181.

Chapter 5

Constraints of HACCP Application on Edible Insect for Food and Feed

Maria João Ramos Fraqueza and
Luis Avelino da Silva Coutinho Patarata

Additional information is available at the end of the chapter

http://dx.doi.org/10.5772/intechopen.69300

Abstract

In a near future there is a need to guaranty food security for approximately more than one billion people worldwide. Beyond the population increase other factors contribute to food and feed insecurity such as climate changes, rising costs of animal protein and consumer demands for protein. Edible insects are pointed out as one alternative as they have always been a part of human diets. The concern to guaranty food security cannot be dissociated from food safety and under the *Codex Alimentarius* principles of food hygiene, insects would be comparable to other types of foods of animal origin. The processing and storage of insects and their products should follow the same health and sanitation regulations as for any other food or feed in order to ensure their microbiological and chemical safety. Allergies induced through insects' ingestion should also be considered. This review aims to identify potential hazards related to edible insects' production and transformation. Preventive measures to their control will be presented considering pre-requirements in their production and transformation. An HACCP plan will be described as a study case in insects' transformation, being discussed all constraints regarding implementation.

Keywords: edible insects, potential hazards, preventive measures, HACCP, safety

1. Introduction

Edible insects have gained interest in recent years all over the world as a solution to mitigate the lack of protein in a near future. With more than 9 billion people worldwide in next 20 years, there is a need to guaranty food security. Beyond the population increase, other factors contribute to food and feed insecurity such as climate changes, rising costs of animal protein, and consumer demands for protein. Alternative solutions to conventional livestock and feed sources urgently need to be found, and edible insects are pointed out as one alternative.

According to the FAO (Food and Agriculture Organization), nearly 2.5 billion humans regularly eat insects in the world. The majority of edible insects are gathered from forest habitats, but in many countries, innovation in mass-rearing systems has begun. Edible insects have always been a part of human diets. They contain high-quality protein, essential amino acids, and vitamins for humans [1].

The concern to guaranty food security cannot be dissociated from food safety, and under the *Codex Alimentarius* principles of food hygiene, insects would be comparable to other types of foods of animal origin [2]. The processing and storage of insects and their products should follow the same health and sanitation regulations as for any other traditional food or feed items in order to ensure food safety. Because of their biological composition, several issues should be considered, such as microbial safety, toxicity, palatability, and the presence of inorganic compounds. Specific health implications should also be observed when insects for feed are reared on waste products such as manure or slaughterhouse waste. Allergies induced through insects' ingestion should be taken into consideration.

This review aims to identify potential hazards related to edible insects' production and transformation. Preventive measures to their control will be presented considering pre-requirements in their production and transformation. A Hazard Analysis Critical Control Points (HACCP) plan will be described as a study case in insects' transformation, being discussed all constraints regarding implementation.

2. Edible insect: intended uses and consumption

Entomophagy is highly rooted in the eating habits of populations from certain regions of the globe [3]. Capture of insects from their natural habitat has been a current practice in African, Asian, and South-American communities [4]. This capture of insects from the nature, without any effort to raise them, can be seen by two completely different points of view. The insects captured in certain ecosystems might result in ecological imbalances with potential negative outcomes [5]. On the contrary, the capture can be strategic to control insect plagues, namely of locusts, resulting in an environmental advantage due to the possibility of reducing insecticides use [6]. Both of those situations are associated with specific populations and connected to the direct capture of insects in the nature. In the so-called developed countries of North hemisphere, the consumption of insects does not have any cultural tradition [7]. Nonetheless, edible insects began to have market penetration in Europe and North America, more as a gourmet product than as a cheap protein source. In fact, edible insects marketed in Europe are very expensive, eventually more expensive than meat [8]. The legal framework of insects and insect-based production in Europe refers to general regulations on food safety, but it can also fall in the category of novel food. In that case, that remains unclear, insects can be considered a novel food, once from the actual regulation *"food consisting of, isolated from or produced from animals or their parts, except for animals obtained by traditional breeding practices which have been used for food production within the EU before 1997 and the food from those animals has a history of safe food use within the Union"* and "*traditional food from a third country (…) with a history of safe food use in a third country*" [9] the classification as novel food depends on the assumption that insects were used or not for food in EU before 1997. Nonetheless, the traditional and long use in certain regions allow to consider edible insects as safe for humans.

In traditional consumption regions, due to insects' nutritional value and consumer demands, a transition has been observed from collection to farm rearing and transition from small domestic productions for self-consumption and local markets to more sophisticated close-cycle farms producing in large scale to a broader market [10]. In order to reduce the production costs and make edible insects price competitive, it is necessary to intensify and automatize the production, since the work cost in Europe and North America is high. Insect farms can be used to produce, virtually, any insect species, provided that it became economically profitable. In an insect farm, all the lifecycle stages should be assured, to guarantee the progeny and continue the production with a new batch [11]. Usually, only a defined stage of the insect's life cycle stage is used for food (**Figure 1**).

There are thousands of insect species that are used as food around the world. Several very complete inventories of edible species insects have been published in [1, 12, 13]. Facing to the growing interest on insects consumption, the European Food Safety Authority (EFSA) promoted, through the Scientific Committee, an evaluation of *the risk profile related to production and consumption of insects as food and feed* [2]. According to this report, there are only few species that have been produced for food with commercial purposes, both inside or outside Europe:

- **Crickets:** *Acheta domesticus, Grillus bimaculatus,* and *Teloegryllus testaceus.*
- **Grasshoppers/locusts:** *Oxya* spp., *Melanoplus* spp., *Hieroglyphus* spp., *Acridia* spp., and *Locusta migratora.*

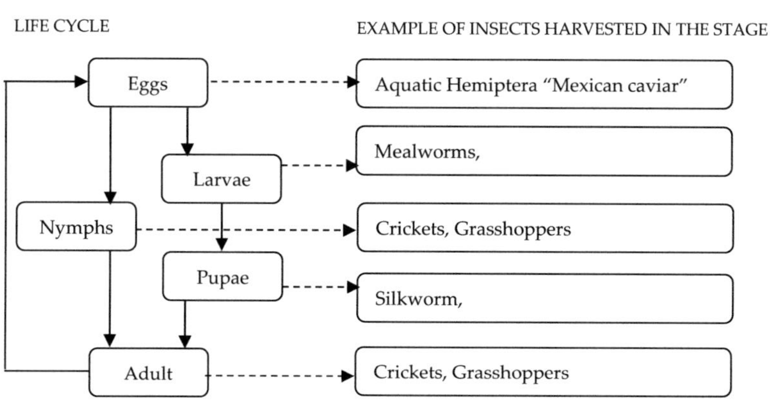

Figure 1. Edible insect's life cycle stages and examples used as food resulting from a specific stage. Adapted from Ref. [8].

- **Mealworms:** *Tenebrio molitor, Alphitobius diaperinus* (lesser mealworm), *Zophobas atratus* (superworm).
- **Silkworm:** *Bombix mori.*

When the production is near to the retail point of consumption, most of these species are harvested and marketed alive or freshly cooked. However, on a larger production, insects must be preserved after harvested.

The culinary use of insects is related to the gastronomic habits of the region and the insect type. They can be fried, sautéed, boiled, roasted, toasted, smoked, and used as ingredient in soups and pizza, among others. Paraphrasing Ref. [14], "*the ultimate goal is to elevate certain edible insects to gourmet food status.*" That perspective of entomophagy is reflected in the number of cooking books available with recipes to prepare insect. Using the keywords "Cookbook insects," Amazon.com retrieves about two dozens of results (accessed March 3, 2017). These insects might have different commercial presentations, namely whole insect, chilled, frozen, dried, snacks, or transformed in flour or pastes to be further processed [13].

3. Food production

The intensive production of insects needs to guarantee food safety, shelf-life and diversifications of the product or new presentations to reach different potential target consumers and overcame the neophobia.

3.1. Process and preservation techniques

Processing of edible insects is now a growing reality in order to accomplish consumer's demands and the business opportunity. Even with a small market in Europe and North America, there are several manufacturers of edible insect's products, namely, whole dehydrated insects to be consumed as snacks or ingredients for culinary preparations, and ground as paste or flour used mostly in cookies, snacks, candy, protein bars, hamburgers, and sausages, among others [12, 13, 15–18].

Processing in industries includes operations that are in the frontier between the farming and the processing. While in hot-blooded farm animals, the step of slaughter is a complex operation, in insects industry, that step is simpler and is usually composed by a **fasting** of about 24 hours to reduce the gut content, since it is not possible to remove the gut in almost all of the edible insects, and the labor costs of that operation would be too high. Before slaughter, insects can be **sieved** to eliminate frass. The **slaughter** is made by reducing the temperature (freezing 24 hours, −18°C) or by heat. It is common in some traditional consumers' regions to cook insects alive, for consumption at home or for selling in local markets. If freezing is chosen, it can be advantageous to improve the operation of **removing legs and wings** [19].

Commercial processing with industrial interests might include three main types of products: dry whole insects, natural, salted or sweetened, or with additional seasonings; pastes obtained by mincing insects, and used to prepare meat-like products, as hamburgers or sausages; flours, used in baking cookies, bread, snacks, and protein bars, among others [13].

The processing steps are basically similar to those used in processing similar products from different raw materials, taking into consideration the specificities of the insects. Once insects have naturally a high microbiota at the beginning of the processing, an initial **cooking** in boiling water during 1–5 min results in a substantial reduction of the microflora. However, it might be associated with nutritional losses [20]. Cooking can be made in a seasoned broth to confer the aimed sensory characteristics of the product [17]. After water drainage, insects are **cooled** and should reach a refrigeration temperature rapidly to avoid the multiplication of survival microorganisms and germination of spores during the cooling. Cooked insects are only pasteurized. Considering its high initial microbiota, they should never be stored at room temperature to avoid spoilage. In products using the **drying** technology, the temperatures used should be adequate to avoid nutritional losses. That operation might be conducted in ovens with temperature varying from 60°C to more than 100°C [21]. During this operation, the insects' water is loss and there is a significant reduction in the microbiota. The hotter is the process, the higher is the microbial reduction, as demonstrated in cooked and dried crickets [22]. Freeze-drying is an interesting alternative to conventional drying, once the water is evaporated directly from the frozen insect, resulting in lesser nutritional and sensory losses, and as all the processes occur at a very low temperature, there is no opportunity for microbial growth [23]. Drying should be controlled to assure that final products have an activity of water enough reduced to inhibit microbial growth. When flours are prepared from dried insects, they are **grinded** to the appropriate size and **sifted**. Further operations made with flours are similar to those using grain flours, having in consideration its technological proprieties [24]. **Minced** products are made from freshly harvested or cooked insects. The raw material is minced to adequate particle size, and the preparation of the product is made, combining it with other ingredients or not, usually to obtain foods similar to those prepared with meat—hamburgers and sausages—aiming to overcome the initial disgusting that the insect itself produces in new consumers [25–27]. Once these products have a high water activity and pH, storage must be done at low temperature, **refrigeration** with a short shelf-life or **freezing**. No matter the processing used, the final product should be correctly **packaged** to avoid recontamination. Considering the high fat levels of most of the edible insects, and particularly the amount of mono and polyunsaturated fatty acids, the package should ideally be made under **modified atmosphere** to avoid or retard lipid oxidation [28]. There are several other processing operations that can be applied to separate fractions with particular nutritional interest [1].

In order to exemplify the processing of edible insects, it is presented as an example of a dry snack prepared from crickets or grasshoppers (**Figure 2(a)**) and flour from mealworms (**Figure 2(b)**).

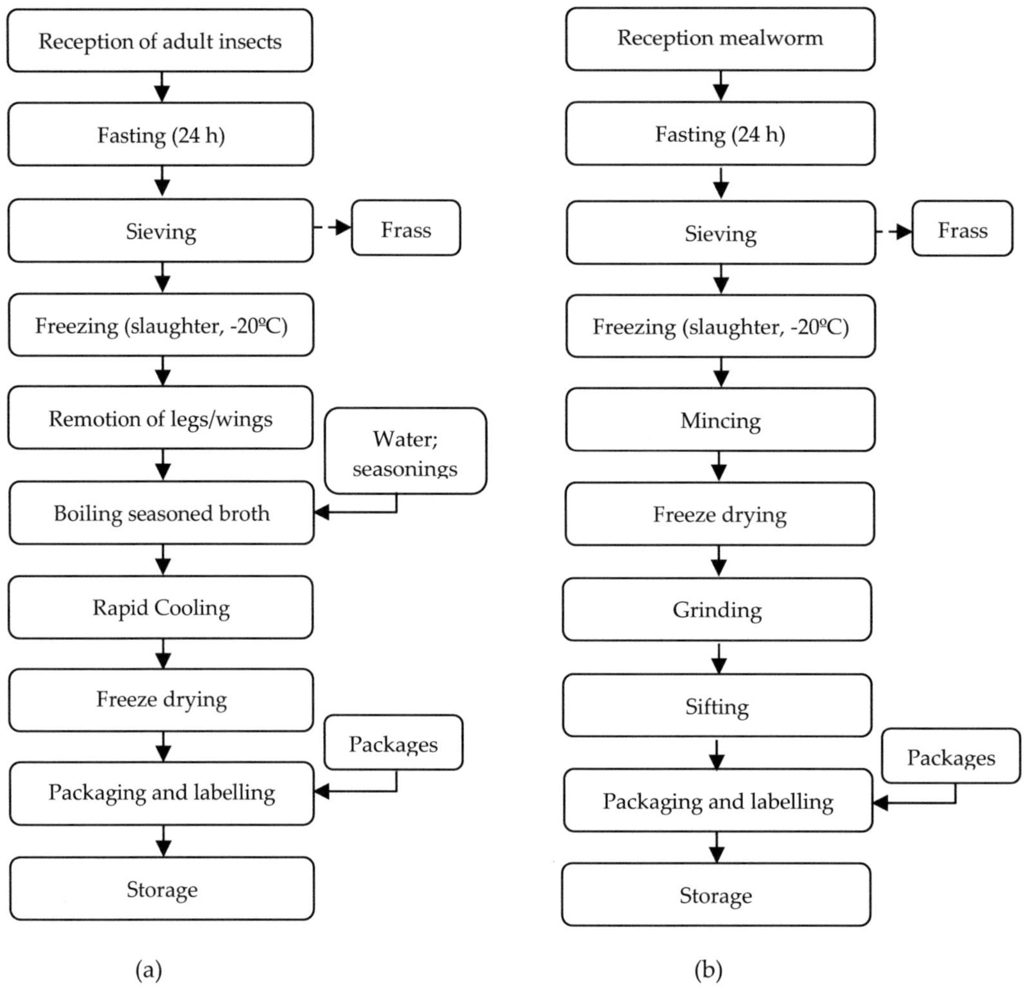

Figure 2. Processing flowchart of (a) cricket/grasshopper dry snacks and (b) mealworm larvae flour. Adapted from Refs. [17, 20, 29].

4. Implementation of an HACCP plan: key to the development of the edible insect sector

The Hazard Analysis Critical Control Points (HACCP) methodology is a systematic tool that identifies specific potential hazards and establishes control systems to ensure the safety of food [30]. The principles preconized on the HACCP methodology are well known, and at food processing level, all the safety management systems were established from *Codex Alimentarius* or on Standards based on it [31, 32]. The identification of biological, chemical, and physical hazards in all materials and process steps is Principle 1. After analysis of the potential hazards identified assessing a consistent risk, according to the preventive measures designed to control hazards, Principle 2 suggests the identification of adequate critical control points (CCPs) in materials and process steps to control the hazards. Principle 3 is the definition of critical limits for each CCP. Principle 4 is monitoring to assure fulfillment of procedures at each stated CCP. When non-conformities from the outlined limits occur at a CCP, Principle 5 states that corrective action should be immediately applied to restore the control. Actions of verification to evaluate compliance of the HACCP plan are stated in Principle 6. All the procedures planed and implemented for monitoring, corrective actions, and verification performed should be recorded, providing an effective demonstration of the system at work in Principle 7. HACCP plan improvement is needed, and its update is based on any output of the system and when any input to the system changes such us premises, layouts, process, or if new legal or trade requirements are introduced and could have impact on safety.The HACCP plan is specific of a certain product, process, and organization. The use of a generic HACCP model to build specific HACCP plans should be taken as an assisting document that always needs creative adaptation and changing. Pre-requisites need to be fulfilled to accomplish with success the implementation of an HACCP plan. They are mandatory for food safety [33].

4.1. Pre-requisites: good farming practices and good practices for processing insect products

In the sector of edible insects', the implementation of preventive measures should be considered at two levels, primary production and processing. The application of HACCP in the primary production is not obligatory in EU, still it is recommended to apply the general principles established in the food safety regulations. On the other hand, insects processing, which includes all the activities after the harvest, namely slaughter, seasoning, cooking, freezing, drying, mincing, grinding, and packaging, among others, should comply with the HACCP methodology [34].

In the **primary production**, the rearing of insects, the general principles of food safety can be applied through an ensemble of good practices that are commonly designated by good hygiene practices (GHP), good agricultural practices (GAP), good veterinary practices (GVP), or to use the vocabulary currently accepted in the HACCP methodology, pre-requisites program (PRP) [35]. Pre-requisite programs include horizontal measures to reduce the odds of hazards introduction in insects' production chain, avoiding biological, chemical, and physical contamination, including cross-contamination, maintaining the hazards probability of occurrence as low as it is possible to achieve [36]. The applicability of HACCP principles to primary production has several constraints, as elucidated in Ref. [37]. These authors propose to use instead the HACCP principles, the preventive measures (pre-requisites) that are grouped in eight points: Environment; Equipment; Personnel; Inputs; Waste; Storage and Transport; Health Status; Recording and Traceability. These are also applied at processing level with particularities related to premises and equipment hygiene and maintenance programs with high relevance to avoid cross contamination and persistence of pathogens and hazardous chemicals [38, 39].

In the **environment** of production, several aspects should be considered. In the planning of farm location, it should be studied to avoid farm implementation in polluted areas that can be responsible for biological or chemical contamination of insects. When in insects' production, it is considered to use low-value lands, and it has been referred to use the natural streams and waste streams [11]. Those practices are not recommended due to the possibility of contamination of insects with several types of hazards. Paraphrasing the Annex III of EU Regulation 767 [40] in Europe, it is prohibited due to the interdiction to use *"all waste obtained from the various phases of the urban, domestic and industrial waste water."* Additionally, it is expected that labor costs for harvesting would be unacceptable [8]. To have a better control on production factors, preventive safety measures, and to reduce the production costs and render the insects' husbandry profitable, the close cycle is recommended [1]. The premises should obey to hygienic design, assuring that contamination is prevented, is built in materials that allow an effective cleaning and disinfection, and avoids pest infestation that might be vehicles of biological hazards [41]. The premises should have adequate sanitary facilities to the personnel. Cleaning and disinfection should be made with a pre-determined frequency and whenever it is necessary. When farms operate with batches in the same life cycle stage, it is common to perform a general cleaning and disinfection in the change of batches. If the farm operates with batches in different stages, cleaning and disinfection must be organized according to the production flow. These practices must be documented, identifying what detergents and disinfectants are used, the concentrations, times of application, as well as the regularity and personnel responsibility [41]. The **equipment** used in insects' husbandry can differ from simple cages in the simplest units, to more complex and sophisticated ones, namely egg separators in reproduction sections, harvest devices, and cold equipment to slaughter the insects. These equipments should comply with general principles of hygiene [35]. **Personnel** should follow personal hygiene practices, use adequate working uniform, and have adequate training and professional attitude. Though insects are phylogenetically very distinct from humans and thus it is highly improbable that any disease is shared between both, they might be vectors of human diseases. Several well-known diseases have a stage of spreading in an insect. These diseases are not probably the major concern, once the insects involved are mostly aphids and mosquitos that are blood suckers and do not have interest for entomophagy. However, insects might be carriers of pathogens that are usually found in warm-blooded animals, like pathogenic Enterobacteriaceae [2]. Thus, the personnel hygiene and the health status of the personnel should be supervised and adequate training should be available to improve the awareness of the workers to the risk factor, they can represent and follow high patterns of good practices [42]. The **inputs** for insects' rearing are probably one of the most sensitive pre-requisite to attend high levels of food safety. **Water and feed** are potential sources of several biological and chemical hazards [37]. Water used at insect farms should be potable to prevent any direct or indirect contamination. Crickets and grasshoppers/locusts are omnivorous, thus they are fed with vegetables and vegetable by-products or formulated feed. To feed mealworms, cereal or cereal by-products or formulated feed is currently used. Other insects have a highly specialized feeding, as silkworms that are nourished only with mulberry leaves [11, 15]. The feed must comply with the safety criteria previewed for any food production animal, in Europe by the Regulation 767 [40]. It is forbidden to use the *"content of digestive and urinary tract of animals slaughtered for human consumption, plant-propagating material, including seeds, treated with plant protection chemical, wood and sawdust treated with preservatives, urban, domestic and industrial waste (solid or liquid), independently of any previous treatment."* Facing these restrictions, several applications of insect production to take advantage of certain sub-products should be carefully considered. Feeding edible insects with catering and domestic leftovers is not recommended, or even allowed by law. Also, the use of animal and human manure is not recommend, due to the high potential that feces have to transport agents responsible for numerous diseases [2], and due to the difficulty in assuring that manure curing destroys a significant part of its potential pathogenic microbiota [43]. When high productivity is aimed, feed should be formulated to fulfill the nutritional needs of the insects. There are several references to the use of formulated feed for chicken [11, 44] or specifically formulated feed to the specific insect [1, 45]. These feeds must obey to the parameters required for any preparation used in animal production. As for industry, a particular selection and control should be implemented for suppliers in order to avoid the entrance of hazards in the production or processing system.

The **waste** management from insect production should include a plan to dispose the debris produced by the insects, namely, its frass, and the unused feed and the substrate used as bed. This waste is usually used as fertilizer for plants [11]. **Storage and transport** should be carefully planned. After harvesting, insects can be transported still alive to be sold directly to the consumer or to the processing units. The good hygiene that is required to the premises should also be applied to vehicles used in transport. Any intermediate storage has to be done in adequate facilities, and when insects are dead, temperature control is obligatory. While the insects' **health status** is not seen as a problem for the safety of the edible insects, that situation might change with the intensification of the production. Intensive production results in a higher probability of spreading diseases for the raising animals [46]. The experience from rearing bees has shown that animal-specific diseases might have negative economic outcomes [47]; thus, if the insects' production takes off for higher levels of intensive production, it is predictable that specific diseases might spread in the farms, and administration of veterinary drugs has to be considered, creating a problem with drug residues similar to that we face actually with meat from conventional husbandry [48, 49]. **Recording** of all procedures and measures that have impact on food safety should be made continuously. These records, as well as a clear identification of production batches, are determinant to ensure that **traceability** is achieved [50]. In case of any incident detected after expedition, the traceability will allow a rapid localization of affected products and an efficient recall.

Standard requirements of hygiene for the production, handling, packaging, storing, and distribution of edible insects and resulting products are mandatory to assure a healthy and wholesome supply of such products and the successful construction and implementation of an HACCP program.

4.2. HACCP model for edible insects processing

4.2.1. Potential hazards identification

The identification of biological, chemical, and physical hazards should be listed for all edible insects used as raw material as well all ingredients or accessory material (primary package) used and process steps related to processing. This hazard identification reveals the main concern of the organization to control them. The increasing interest in insects for food and feed does not correspond to the available literature that could be used to identify potential hazards related to those edible insects. More studies will be necessary to understand and validate several clues regarding identification of potential hazards or even emerging potential hazards. The European

Food Safety Agency (EFSA) has been trying to do the identification of hazards regarding edible insects; however, many gaps still exist regarding the occurrence of human and animal pathogens in insects processed for food and feed [2]. That European organism considered that the risk associated with edible insect consumption is similar to other food consumption.

Regarding potential microbial hazards, it is known that insects have their own microbiota and can serve as vectors for microorganisms pathogenic to humans [51]. Intrinsic microbiota of insects includes Enterobacteriaceae (*Proteus, Escherichia*), *Pseudomonas, Staphylococcus, Streptococcus, Bacillus, Micrococcus, Lactobacillus,* and *Acinetobacter* [52–54]. These families/genera can be related with specific pathogens for humans, while others are considered non pathogens in healthy people. In **Table 1** are listed **microbiological hazards** on edible insects

Insects	Potential hazards	Causes	Scientific evidence	Preventive measures
Tenebrio molitor (yellow meal beetle) *Schistocerca gregaria* (desert locust) *Bombyx mori* (silkmoth) *Acheta domesticus* (cricket) *Locusta migratoria* (whole locust),	*Salmonella, E. coli, Staphylococcus aureus, E. faecalis, E. faecium, Aeromonas hydrophila Bacillus cereus, Clostridium perfringens, Cl. septicum, Clostridium difficile, Cl. sporogenes, Listeria* spp.	Rearing conditions, farming practices, handling	[51, 62–65]	PRP
Locusta migratoria (migratory locust)	Vesicular stomatitis virus		[19, 66]	
Imbrasia bellina/ Gonimbrasia bellina, (mopani worm, emperor moth) *Atta laevigata* (leaf cutter ants)	*Aspergillus fumigatus, Aspergillus sclerotiorum, Penicillium, Fusarium,* Cladosporium, and Phycomycetes		[4, 59, 67, 68]	
Aquatic insects Insect larva	Cercaria and metacercaria Nematodes *Dicrocoelium dendriticum ; Plagiorchis*		[21, 69]	
Acheta domesticus (small crickets) *Locusta migratoria* (locusts) *Hyboschema contractum* (Rhino beetles) *Gryllotalpidae* (mole crickets)	Antibiotic-resistant genes E.g., *tet(M) tet(O) tet(S) tet(K), erm(B), blaZ*	Farming practices, improper use of antibiotics, fail on biosecurity	[70]	PRP Farm biosecurity
Edible insects in general	Mechanical or biological vectors of prions	Rearing conditions, farming practices, handling	[71–74]	PRP

Table 1. Identification of biological hazards associated with raw materials/ingredients reported, principal causes, and preventive measures.

and the respective scientific evidence. *Campylobacter*, verotoxigenic *Escherichia coli*, *Salmonella*, and *Listeria monocytogenes* may be present in non-processed insects and their occurrence should be assessed. The detection of some pathogens in some edible insect has been reported (**Table 1**), while the presence of others was not yet demonstrated. The prevalence of some of these pathogens is lower when compared with other sources of animal protein, for example, *Campylobacter* does not replicate in the intestinal tract of the insects [55–57].

Insects can have virus pathogenic to their self and are considered virus vectors implicated in many plant and animal diseases. The Rhabdovirus causing vesicular stomatitis in animals and humans has been reported on edible insects (**Table 1**). However, there is also a lack of information related to the likelihood of human viruses such as norovirus, rotavirus, and Hepatitis E and A, being passively transferred from feedstock through residual insect gut contents [2].

Insects can harbor protozoa implicated in animal and human diseases. Several human parasitic diseases implicate insects as infected vectors for the transmission, for example, of *Trypanosoma* and *Leishmania*. The parasitic foodborne diseases related to edible insects are not well documented. It was reported that the trematode *Dicrocoelium dendriticum* (family *Dicrocoeliidae*) is a parasitic zoonotic agent that potentially infects humans through insect consumption. There are evidences suggesting the possible foodborne transmission of parasites (trematodes) belonging to the families *Lecithodendridae* and *Plagiorchidae* [58]. Despite the reported occurrence of parasites in the insects and the relationship between sporadic human parasitic disease and insect consumption, there are no data on the occurrence of parasites in farmed insects. Insects may also be carriers of fungi and yeasts with potential hazard to animals and humans. Yeasts and fungi were found in considerable amounts in fresh, freeze-dried as well as in frozen insects (*T. molitor* and *Locusta migratoria*) in Ref. [2].

The importance of proper processing, handling, drying, and storage was stressed regarding aflatoxins in some commercial lots of mopane worms (*Gonimbrasia belina*: *Saturniidae*) [4]. From the same species, dried under laboratory conditions, some fungi were isolated (*Aspergillus* spp. and *Penicillium* spp.), among which are mycotoxigenic species [59]. In general, any hazard from fungi associated with insects produced for food and feed or introduced during farming, processing, and storage could be mitigated by hygienic measures in the entire production chain.

Three main issues with prion-related risks from insects were reported by EFSA [2]: insect-specific prions, insects as mechanical vectors of animal/human prions, and insects as biological vectors of prions (i.e., involving replication of animal/human prions within insects).

The **chemical hazards** identified in **Table 2** could result from substances synthetized by the insect itself or substances accumulated by the insect from its environment or feed. Not all insect species are therefore edible or may be edible depending on the stage of the life cycle [19]. Toxins and antinutritional factors should be assessed in order to select insects for feed and food. The main chemical hazards present on edible insects are related with farming practices and rearing conditions that should comply with the applicable food safety regulations. Edible insects entering in the production of food and food ingredients should be kept in such a way to prevent or minimize the accumulation of externally introduced toxins, drugs, or antinutrients. Chitin, a constituent of the insect exoskeleton, and chitosan, one of its derivatives, should be regarded as antinutritional factors and also as a potential allergenic substance with complex effect on the immune system. Another serious concern related to edible insects is related to allergic reactions induced by substances present on edible insects; however, there are a very limited number of studies available [21]. Allergens identified and listed (**Table 2**) should be indicated on the label of edible insect's products.

The use of veterinary drugs should be planned in insect farms to reduce the mortality, especially associated with bacterial or parasitic infections. The information related to the presence of veterinary drug residues in insect is very scarce. The use of chloramphenicol, a broad-spectrum antibiotic prohibited in animal production [60], was referred to treat a disease in silkworms (*Bombyx mori*). Other biocides should be considered as potential hazards since the control of insects' diseases in intensive production will probably be needed.

Insects might accumulate hazardous chemicals, including heavy metals, dioxins, and flame retardants. Data on hazardous chemicals in reared insects and in insect-based food are scarce. Recently, chemical hazards on edible insect and their detection were shown [61]. Apart from the environmental contaminations and the possibility for several chemicals to accumulate in the farmed insect, the levels of contamination were relatively low and concentrations were similar or lower than those measured in commonly eaten animal products, such as meat, fish, and eggs.

Insects have a high-quality amino acid profile with high contents of phenylalanine and tyrosine [8]. These amino acids could give rise to the formation of biogenic amines tyramine and histidine associated with aminogenic microorganism and inadequate conditions of storage. Also, contamination with molds could increment the potential hazard micotoxins (Aflatoxins, Beauvericin; Enniatin A; Enniatin A1) described by several authors [19].

Physical hazards are dense contaminants or foreign bodies that might cause injury to the consumer. Edible insects are not especially prone to be a vector of physical hazards. During edible insects' processing, a recontamination by foreign bodies from the process (metal and plastic), as with any other processed food, could occur. Generally, all the edible insects might have hard parts: elytra, rostrums, and wings; and leaves and soil could be considered hazards, particularly for unused consumers. The prevention of its occurrence can be achieved by a strict compliance with the pre-requisites plan.

Tables 1 and **2** list the main biological and chemical hazards, respectively, identified in edible insects. Physical hazards are not listed once this production does not have particular physical hazards, beyond those previously described. **Table 3** presents the potential hazards associated with several processing operations that can be used in insects processing, namely, freezing, freeze-drying, dehydration, boiling, deep-frying, toasting, cooling, acidification, packaging, and end product storage, as well as the preventive measures that can be applied to prevent its occurrence in the final product. In the tables, PRP refers to the Pre-Requisite Plan, as discussed in item 4.1.

Hazard analysis and assessment performed by the food industry are qualitative using a bidimensional model based on hazards severity and probability of occurrence [30], but a quantitative analysis can only be performed when detailed studies based on knowledge of

Insects	Potential hazards	Causes	Scientific evidence	Preventive measures
Oecophylla smaragdina (Hymenoptera: Formicidae); Odontotermes sp. *(Isoptera:Termitidae); Coptotermes gestroi* (soldier termites, *Rhinotermitideae*); *Cirina forda* (Lepidoptera: Saturniidae)	*Antinutritional factors:* phytic acid, oxalates. hydrocyanic acid, tannins, thiaminase	Species-specific	[75–77]	Labeling
Bombyx mori (silkworm pupae); *Ophiocordyceps sinensis* (caterpillar fungus), *Rhynchophorus ferrugineus* (red palm weevil), *Tenebrio molitor* (yellow meal beetle) *Locusta migratoria* (locust)	Allergens: myosin, troponin, a-amylase, tropomyosin, arginine Kinase, hémocyanine, Hexamérine, -amylase, Arginine kinase, chitinase, glutathion-S-transférase, triose, phosphate isomérase, trypsine, Chitin, pollen, Histamine		[23, 69, 78, 79]	
Lytta vesicatoria (Spanish fly) Tenebrionidae (darkling beetles) moth species (*Zygaena*)	*Toxic substances*: cantharidin, amonoterpene (2,6-dimethyl-4,10-dioxatricyclo-[5.2.1.02,6]decane-3,5-dione), quinones, and alkanes cyanogenic glycosides		[19]	PRP selection of edible insects
All edible insects	Micotoxins: Aflatoxins, Beauvericin; Enniatin A; Enniatin A1	Rearing conditions, farming practices,	[19]	PRP
Tenebrio molitor (yellow meal beetle)	Pesticides (ex. clopyralid, benzoquinons), persistent organic pollutants	Farming practices, improper use of pesticides	[80]	
All edible insects	Dioxins, polychlorinated non-ortho and mono-ortho biphenyls (dioxin like PCBs), organochlorine compounds (OCPs), poly- brominated diphenyl ethers (PBDEs),		[61, 81]	
All edible insects	Heavy metals: cadmium, lead, arsenic, zinc, copper	Farming practices, improper use of antibiotics, and fail on biosecurity	[61, 82–85]	
Bombyx mori (silkworms) *M. domestica*	Veterinary drug residues (ex. chloramphenicol, 4,4'-dinitrocarbanilide (nicarbazin))		[86, 87]	

Table 2. Identification of chemical hazards associated with raw materials/ingredients reported, principal causes, and preventive measures.

Processing steps	Potential hazards	Causes	Preventive measures
Reception of raw material	All hazards mentioned for all edible insects	Fails on temperature; without safety specification for purchase	Selection and control of suppliers, definition of safety criteria for purchase, temperature of reception
Reception of seasoning (herbs and spices) and additives	**Biological:** Pathogenic microorganisms: *Salmonella, L. monocytogenes; S. aureus, E. coli* (VTEC)*Clostridium perfringens spores;* and *Aspergillus flavus* **Chemical**: Pesticides (aldrin, linden, etion), herbicides, dioxins, heavy metals (selenium), additives legal requirement for toxic dose, micotoxins (Aflatoxins) **Physical**: Soil, stones, wood, and plastic fragments		
Freezing (slaughter)	**Biological:** Pathogenic microorganisms multiplication present on raw material **Chemical**: None **Physical**: None	Fails on temperature, improper hygiene, and maintenance of equipment	Correct temperature and time, preventive maintenance plan for equipment
Bleaching	**Biological:** no inactivation of pathogenic microorganisms present on raw material due to fails on temperature/time **Chemical**: Heavy metals **Physical**: None	Improper maintenance of equipment, incorrect binomial temperature, and time for bleaching, improper water	Correct temperature and time, preventive maintenance plan for equipment, potable water
Boiling	**Biological:** no inactivation of pathogenic microorganisms present on raw material **Chemical**: Heavy metals **Physical**: None	Improper maintenance of equipment, incorrect binomial temperature and time for boiling, improper water	
Rapid cooling	**Biological:** recontamination with pathogenic microorganisms (spores germination) and growth (*Cl. perfringens* and other pathogenic *Bacillaceae, L. monocytogenes, Salmonella*) **Chemical**: Heavy metals **Physical**: None	Fails on temperature, long time for cooling, improper hygiene, and maintenance of equipment, improper water	Corrected temperature and time for cooling, GHP, potable water, hygiene program for equipment and premises, preventive maintenance plan for equipment and premises
Storage under Refrigeration	**Biological:** Pathogenic microorganisms multiplication present on raw material **Chemical**: Histamine, Micotoxins: Aflatoxins, Beauvericin; Enniatin A; Enniatin A1 **Physical**: None	Fails on temperature, incorrect relative humidity, improper practices of hygiene and maintenance of equipment, long time of storage	Corrected temperature and relative humidity, GHP, hygiene program for equipment and premises, preventive maintenance plan for equipment and premises

Processing steps	Potential hazards	Causes	Preventive measures
Mincing	**Biological:** Pathogenic microorganisms multiplication/ contamination **Chemical**: Heavy metals from water **Physical**: Metals particles		GHP, potable water, hygiene program for equipment and premises, preventive maintenance plan for equipment and premises
Freeze drying	**Biological:** Pathogenic microorganisms multiplication present on raw material **Chemical**: None **Physical**: None	Improper maintenance of equipment, incorrect relative humidity, fail on aw reduction	Corrected air velocity and relative humidity, preventive maintenance plan for equipment and premises
Grinding	**Biological:** Pathogenic microorganisms multiplication **Chemical**: None **Physical**: Metals particles	Improper practices of hygiene and maintenance of equipment, improper practices of processing with long time of grinding	Preventive maintenance plan for equipment and premises
Packaging finished products and labeling	**Biological:** Pathogenic microorganisms contamination (*Aspergillus fumigatus, Aspergillus sclerotiorum, Penicillium, Fusarium; L. monocytogenes; Salmonella*) **Chemical**: Packaging migration contaminants: ink, bisphenol A and phthalates, Allergens not identified in the label **Physical**: Metals	Improper packaging material, allergens, improper sealing, incorrect labeling	Selection and control of suppliers, GHP, GMP, control of good sealing, detection of metals on real time, preventive maintenance plan for equipment, correct labeling, GMP
Storage end product	**Biological**: Recontamination with pathogenic microorganism and growth of *Aspergillus fumigatus, Aspergillus sclerotiorum, Penicillium, Fusarium* **Chemical**: Tiramine, Histamine, Micotoxins: Aflatoxins, Beauvericin; Enniatin A; Enniatin A1 **Physical**: None	Storage conditions Fails on temperature/ humidity Long time of storage, fails on GHP and GMP	GHP, GMP Correct time/temperature/ humidity, shelf-life validation

Table 3. Identification of potential hazards associated with the processing operations potentially used in insect processing, principal causes, and preventive measures.

likelihood of exposure to a specific hazard are available. The analysis of the hazards identified on edible insects should be performed; however, the occurrence of several hazards is not well documented. This could be a major constraint in the planning of a safety management system for an organization related to edible insects processing, since the hazards analysis and risk might not be adequately established. Preventive measures are actions taken to control the hazards. These include physical, chemical, or biological factors or other hurdles required to control a hazard likely to occur at particular stages of the processing of insects' products. The occurrence of hazards identified in an organization depends on preventive measures already implemented. Several preventive measures may be required to control a specific hazard. Likewise, there are hazards that can be effectively controlled by one single preventive measure. The control of some preventive measures designed to eliminate or reduce potential hazards may be assured by validated pre-requisites programs (PRP), also named Good Hygiene and Good Processing Practices (GHP/GMP). From the hazards analysis, according to their occurrence and severity, only some corresponding to a high or moderate risk level will be questioned for the identification of critical control points.

4.3. Identification of critical control points on edible insects processing

A critical control point (CCP) is a step or procedure where control can be applied and a hazard can be prevented, eliminated, or reduced to acceptable levels. The determination of a CCP in a process should use the safety knowledge about edible insects, only for real and likely occurring hazards and where preventive measures are available for their control. The *Decision Tree* for raw materials and processing steps is an algorithm to help finding the adequate CCPs for each considered hazard [32]. Critical control points require strict monitoring, which implies costs. So, CCP should be in the minimum number if it is possible, no more than three or four in each plan. After identification of all CCPs, it is needed to decide how their control will be done and establish the criteria for acceptance. The absolute tolerance at a CCP is known as a critical limit [88]. In **Table 4**, some of the (potential) identified CCPs on insects processing are summarized. An effective monitoring operation of CCPs is fundamental to assure product safety and should be preconized by defining critical limits of parameters that must be under control.

CCP and location *(Principle 2)*	**Parameters to establish critical limits** *(Principle 3)*	**Monitoring procedures and frequency** *(Principle 4)*	**Corrective actions** *(Principle 5)*	**Verification procedures** *(Principle 6)*
Drying or freeze drying	Relative humidity/air velocity/temperature duration according to the process preconized Weight loss (related with aw). aw < 0.60	Measure humidity/ air velocity/ temperature/time of the operation; Take a sample to measure the weight loss to achieve the desired aw per batch	Increase the drying period; reprocess	Take a sample to measure water activity in 5% of the batches produced according to a plan of sampling
Packaging	0% of metal particles 0% of fail on sealing package	100% packages screened on the metal detector 5% batch package sealing control	Reject packages with metal hazards and not sealed; Review the equipment	Calibration of metal detector with control packages with known metal samples size
Labelling	Allergenic hazard presence should be communicated and be clearly identified in all the labeling	Trained workers should inspect all the packages for the presence of label	Packages without label with incorrect information should be drawn from the line to be labeled again	A random sample of packages should be drawn to check for integrity and readability of the label

Table 4. Example of an HACCP plan with possible CCP for mealworm flour.

4.4. Critical limits establishment for CCP and monitoring actions

Critical limits must be associated with a measurable factor that can be routinely monitored according to a fixed schedule. These critical limits are established based on published data (scientific literature, in-house and supplier records, and regulatory guidelines), experimental data, expert advice, and mathematical modeling [32] and designated to be monitored at CCPs. Monitoring procedures assure that they have been respected. Regarding some technological parameters associated with the manufacture of edible insect products, the available information is scarce. The steps of edible insects' processing for flour or snacks (**Figure 2**) relate with preventive measures that control identified potential hazards and are recognized CCPs for several technological processes. Chilling, freezing, and freeze-drying or drying will have an impact on pathogenic microbiota control, avoiding its multiplication, while blenching, boiling, and other thermic treatment or culinary treatment will be effective and destructive against pathogenic microbiota. All these steps are to be controlled by objective parameters that allow to check on-line all these CCPs, giving the opportunity of immediate correction or corrective measures if a fail is detected. The monitoring frequency depends on the nature of the CCP, the type of monitoring procedure, and the amount of production [88]. The validation of the binomial temperature/time for an effective lethal effect of microorganisms should be considered in practice, being considered a real evidence of the control.

Some chemical hazards identified (mycotoxins and biogenic amines) at processing steps such as storage can be avoided in controlling the producing microorganisms. Aminogenic microorganisms can be avoided when good hygiene practices are effectively implemented. For flour production, some bacteriostatic additives might be introduced, as well as antioxidants. If chemical additives are used, it should be considered as a potential hazard and be controlled at the weighing. This step can be considered a CCP or eventually a good practice.

Monitoring data prove that the process is under control and provide a pool of data that, after trend analysis, give outputs to improve the implemented system. Particularly if the data correspond to fails occurring on the implemented safety system, the cause should be analyzed to a better establishment of corrective actions.

4.5. Corrective actions

The corrective actions are those that are previewed to implement when a deviation occurs, detected by the CCP monitoring, and include those related to the adjustment of the process to bring it back under control and those related to the amount of product that might not be complying with the safety requirements. These actions can include the segregation of any suspect product and holding it for the period of time needed to study the risk it might represent. All this information will lead to different decisions: rejection and destruction of product, product rework, and product release. All these actions must be kept on record. In **Table 4**, some examples related to corrective actions in a dry freezing process for insect flour process are given.

4.6. Verification actions

The HACCP plan must be revised when any modification in production or equipment occurs. The HACCP team is responsible for the revision based on internal and external audit reports, records of corrective actions, and client complaints. Revision will contribute to a continual improvement of the plan.

5. Concluding remarks

Production of insects and insect-based products is a current practice in several regions of the globe with rooted habits of its consumption. In Europe and North America, these products began to explore a potential market of alternative foods. Safety issues related to insects' rearing and processing are a major concern, once due to their composition they are prone to be contaminated with pathogens and hazardous chemicals. Application of HACCP to the production and processing should be performed as for any other animal products used for human consumption. In the primary production of insects, the implementation of a robust pre-requisites plan will guarantee the reduction of the likelihood of incidents involving spreading microorganisms responsible for foodborne diseases or chemical compounds harmful for the consumer. In the processing of edible insects, once the operations used are similar to those used in other industries, the safety concerns are similar to those. In both primary production and processing, the rigorous application of HACCP principles relies on a deep knowledge on the hazards associated with the products and processes. Notwithstanding the generous amount of scientific information published in the last few years on the potential hazards associated with edible insects, the major constraint that we still face is the lack of solid information, related to the occurrence of biological and chemical hazards, how can that situation changes in contexts of production intensification, as well as surveillance or epidemiologic data on foodborne diseases having edible insects as vehicle of the hazard. That information, which is still scarce, will allow the food safety team of the insect industry to conduct an adequate and correctly supported hazard analysis that is the fundamental to the success of HACCP plan and its implementation. Additionally, the direct transposition of critical control points and respective critical limits, particularly those related to biologic hazards, from other industries to the insect processing might also have some risks, once the behavior of pathogens might be different, due to substantial differences in its composition. Thus, validation of control measures considered in CCP should be seriously considered in the insect industry.

The expected growth of insect business has several hurdles to overcome. Probably, the acceptance by new consumers is the biggest one. The rigorous compliance with the food safety rules, through the application of HACCP, will result in safe products, allowing the consumer to profit from its excellent nutritional value and from the sensorial experience.

Acknowledgements

The authors wish to thank the Centro de Investigação Interdisciplinar em Sanidade Animal (CIISA) for the financial support from the *Fundação para a Ciência e a Tecnologia* (FCT) through Project UID/CVT/00276/2013.

Author details

Maria João Ramos Fraqueza[1]* and Luis Avelino da Silva Coutinho Patarata[2]

*Address all correspondence to: mjoaofraqueza@fmv.ulisboa.pt

1 CIISA, Faculty of Veterinary Medicine, University of Lisbon, Lisbon, Portugal

2 CECAV, Universidade de Trás-os-Montes e Alto Douro, Vila Real, Portugal

References

[1] van Huis A, Van Itterbeeck J, Klunder H, Mertens E, Halloran A, Muir G, et al. Edible Insects: Future Prospects for Food and Feed Security. Rome: FAO; 2013. p. 201

[2] EFSA. Risk profile related to production and consumption of insects as food and feed EFSA Scientific Committee. EFSA Journal. 2015;**13**(October):1-60

[3] Shelomi M. The meat of affliction: Insects and the future of food as seen in expo. Trends in Food Science & Technology. [Internet]. 2016;**56**:175-179. Available from: http://dx.doi.org/10.1016/j.tifs.2016.08.004

[4] Schabel H. Forest insects as food: A global review. In: FAO, editor. Forest Insects as Food: Humans Bite Back. Bangkok; 2010. pp. 37-64

[5] Halloran A, Vantomme P. The Contribution of Insects to Food Security, Livelihoods and the Environment [Internet]. FAO. 2017. Available from: http://www.fao.org/docrep/018/i3264e/i3264e00.pdf [Accessed: January 8, 2017]

[6] Cerritos R, Cano-Santana Z. Harvesting grasshoppers *Sphenarium purpurascens* in Mexico for human consumption: A comparison with insecticidal control for managing pest outbreaks. Crop Protection. 2008;**27**(3-5):473-480

[7] Deroy O, Reade B, Spence C. The insectivore's dilemma, and how to take the West out of it. Food Quality and Preference. 2015;**44**:44-55

[8] Rumpold BA, Schlüter OK. Potential and challenges of insects as an innovative source for food and feed production. Innovative Food Science and Emerging Technologies [Internet]. 2013;**17**:1-11. Available from: http://dx.doi.org/10.1016/j.ifset.2012.11.005

[9] Regulation (EU) 2015/2283 of the European Parliament and of the Council of 25 November 2015 on novel foods. Official Journal of the European Union. 2015;**11.12.2015**: L327/1

[10] Halloran A, Roos N, Flore R, Hanboonsong Y. The development of the edible cricket industry in Thailand. Journal of Insects as Food and Feed. 2016;**2**:91-100

[11] Halloran A, Bruun S. Life cycle assessment of edible insects for food protein: A review. Agronomy for Sustainable Development [Internet]. 2016;**36**:1-13. Available from: http://dx.doi.org/10.1007/s13593-016-0392-8

[12] Nowak V, Persijn D, Rittenschober D, Charrondiere UR. Review of food composition data for edible insects. Food Chemistry. 2016;**193**:39-46

[13] Sun-Waterhouse D, Waterhouse GIN, You L, Zhang J, Liu Y, Ma L. Transforming insect biomass into consumer wellness foods: A review. FRIN [Internet]. 2016;**89**:129-151. Available from: http://dx.doi.org/10.1016/j.foodres.2016.10.001

[14] Johnson D. The contribution of edible forest insects to human nutrition and to forest management. In: Forest Insects as Food: Humans Bite Back. Bangkok: FAO; 2010. pp. 4-22

[15] Kim H, Setyabrata D, Jae Y, Jones OG, Brad YH. Pre-treated mealworm larvae and silkworm pupae as a novel protein ingredient in emulsion sausages. Innovative Food Science and Emerging Technologies [Internet]. 2016;**38**:116-123. Available from: http://dx.doi.org/10.1016/j.ifset.2016.09.023

[16] Stoops J, Vandeweyer D, Crauwels S, Verreth C, Boeckx H, Borght M Van Der, et al. Minced meat-like products from mealworm larvae (*Tenebrio molitor* and *Alphitobius diaperinus*): Microbial dynamics during production and storage. Innovative Food Science and Emerging Technologies [Internet]. 2017;**41**:1-9. Available from: http://dx.doi.org/10.1016/j.ifset.2017.02.001

[17] Thailand Unique [Internet]. 2017. Available from: www.thailandunique.com [Accessed: February 10, 2017]

[18] Entomarket [Internet]. 2017. Available from: www.edibleinsects.com [Accessed: February 10, 2017]

[19] Schlüter O, Rumpold B, Holzhauser T, Roth A, Vogel RF, Quasigroch W, et al. Safety aspects of the production of foods and food ingredients from insects. Molecular Nutrition & Food Research [Internet]. 2016;**0**:1-16. Available from: http://onlinelibrary.wiley.com/doi/10.1002/mnfr.201600520/full%5Cnwww.mnf-journal.com

[20] van der Spiegel M, Noordam MY, van der Fels-Klerx HJ. Safety of novel protein sources (insects, microalgae, seaweed, duckweed, and rapeseed) and legislative aspects for their application in food and feed production. Comprehensive Reviews in Food Science and Food Safety. 2013;**12**(6):662-678

[21] ANSES (French Agency for Food E and OH and S). Opinion on the use of Insects as Food and Feed and the Review of Scientific Knowledge on the Health Risks Related to the Consumption of Insects [Internet]. 2015. Available from: https://www.anses.fr/fr/system/files/BIORISK2014sa0153EN.pdf [Accessed: January 7, 2017]

[22] Grabowski NT. Microbiology of cooked and dried edible Mediterranean field crickets (*Gryllus bimaculatus*) and superworms (*Zophobas atratus*) submitted to four different heating treatments. Food Science and Technology International. 2017;**23**:17-23

[23] Han S, Lee B, Jung K, Yu H, Yun E, Sam J, et al. Safety assessment of freeze-dried powdered Tenebrio molitor larvae (yellow mealworm) as novel food source: Evaluation of 90-day toxicity in Sprague-Dawley rats. Regulatory Toxicology and Pharmacology [Internet]. 2016;**77**:206-212. Available from: http://dx.doi.org/10.1016/j.yrtph.2016.03.006

[24] Kinyuru JN, Kenji GM. Effect of processing methods on the in vitro protein digestibility and vitamin content of edible winged termite (*Macrotermes subhylanus*) and grasshopper (*Ruspolia differens*). Food and Bioprocess Technology. 2010;**3**:778-782

[25] Tan HSG, Fischer ARH, Tinchan P, Stieger M, Steenbekkers LPA, van Trijp HCM. Insects as food: Exploring cultural exposure and individual experience as determinants of acceptance. Food Quality and Preference. 2015;**42**:78-89

[26] Shan H, Tan G, Fischer ARH, Van Trijp HCM, Stieger M. Tasty but nasty? Exploring the role of sensory-liking and food appropriateness in the willingness to eat unusual novel foods like insects. Food Quality and Preference [Internet]. 2016;**48**:293-302. Available from: http://dx.doi.org/10.1016/j.foodqual.2015.11.001

[27] Hartmann C, Siegrist M. Becoming an insectivore: Results of an experiment. Food Quality and Preference [Internet]. 2016;**51**:118-122. Available from: http://dx.doi.org/10.1016/j.foodqual.2016.03.003

[28] Usub T, Lertsatitthankorn C, Poomsa-ad N. Food and bioproducts processing. Food and Bioproducts Processing [Internet]. 2009;**88**(2-3):149-160. Available from: http://dx.doi.org/10.1016/j.fbp.2009.04.002

[29] Bußler S, Rumpold BA, Fröhling A, Jander E, Rawel HM, Schlüter OK. Cold atmospheric pressure plasma processing of insect flour from Tenebrio molitor: Impact on microbial load and quality attributes in comparison to dry heat treatment. Innovative Food Science and Emerging Technologies [Internet]. 2016;**36**:277-286. Available from: http://dx.doi.org/10.1016/j.ifset.2016.07.002

[30] Codex Alimentarius. Recommended International Code of Practice General Principles of Food Hygiene; WHO & FAO. Rome, Italy. 2003. p. 31

[31] NACMCF (National Advisory Committee on Microbiological Criteria for Foods). HAZARD analysis and critical control point principles and application guidelines. Journal of Food Protection. 1997;**61**:762-775

[32] Mortimore S, Wallace C. HACCP—A Practical Approach. New York: Springer-Verlag; 2013. p. 475

[33] Arvanitoyannis I, Kassaveti A. HACCP and ISO 22000—A comparison of the two systems. In: Arvanitoyannis I, editor. HACCP and ISO 22000—Application to Foods of Animal Origin. Chichester, UK: John Wiley & Sons; 2009. pp. 3-45

[34] Regulation (EC) No. 852/2004 of the European Parliament and of the Council of 29 April 2004 on the hygiene of foodstuffs. Official Journal of the European Union. 2004; **2004-04-30**:L139/1

[35] Codex Alimentarius. Animal Food Production. Rome: WHO/FAO; 2009. p. 246

[36] ISO 22000. Food Safety Management Systems—Requirements for Any Organization in the Food Chain. International Organization for Standardization; Genève, Switzerland. 2005. 53 p

[37] Cerf O, Donnat E, Working H. Application of hazard analysis—Critical control point (HACCP) principles to primary production: What is feasible and desirable? Food Control. 2011;**22**:1839-1843

[38] Henriques A, Telo Da Gama L, Fraqueza M. Assessing Listeria monocytogenes presence in Portuguese ready-to-eat meat processing industries based on hygienic and safety. Food Research International. 2014;:81-88

[39] Henriques A, Telo Da Gama L, Fraqueza M. Tracking *Listeria monocytogenes* contamination and virulence in the ready-to-eat meat-based food products industry according to the hygiene level. International Journal of Food Microbiology. 2017;**242**:101-106

[40] Regulation (EC) No. 767/2009 of the European Parliament and of the Council of 13 July 2009 on the placing on the market and use of feed. Official Journal of the European Union. 2009;**2009-09-01**:L229/1

[41] Copa, Cocega. EU Guide to Good Hygiene Practice (GGHP) for the Primary Production of Foodstuffs [Internet]. Brussels: European Farmers European Agri-Cooperatives; 2016. p. 34. Available from: http://www.copa-cogeca.eu/ [Accessed: December 12, 2016]

[42] Ko W. Evaluating food safety perceptions and practices for agricultural food handler. Food Control [Internet]. 2010;**21**(4):450-455. Available from: http://dx.doi.org/10.1016/j.foodcont.2009.07.005

[43] Sahlström L. A review of survival of pathogenic bacteria in organic waste used in biogas plants. Bioresource Technology. 2003;**87**:161-166

[44] Hanboonsong Y. Edible insects and associated food habits in Thailand. In: FAO, editor. Forest Insects as Food: Humans Bite Back. Bangkok; 2010. pp. 173-182

[45] van Zyl C. Cost-effective culturing of *Galleria mellonella* and *Tenebrio molitor* and entomopathogenic nematode production in various hosts. African Entomology. 2015; **23**:361-375

[46] Belluco S, Losasso C, Maggioletti M, Alonzi CC, Paoletti MG, Ricci A. Edible insects in a food safety and nutritional perspective: A critical review. Comprehensive reviews in food science and technology. 2013;**12**:296-313

[47] Chantawannakul P, de Guzman LI, Li J, Williams GR. Parasites, pathogens, and pests of honeybees in Asia. Apidologie. 2016;**47**(3):301-324

[48] Niewold TA, van Essen GJ, Nabuurs MJ, Stockhofe-Zurwieden N, van der Meulen J. A review of porcine pathophysiology: A different approach to disease. The Veterinary Quarterly. 2000;**22**(4):209-212

[49] Fraqueza MJ, Martins A, Borges AC, Fernandes MH, Fernandes MJ, Vaz Y, et al. Antimicrobial resistance among Campylobacter spp. Strains isolated from different poultry production systems at slaughterhouse level. Poultry Science. 2014;**93**(6):1578-1586

[50] Bosona T, Gebresenbet G. Food traceability as an integral part of logistics management in food and agricultural supply chain. Food Control [Internet]. 2013;**33**(1):32-48. Available from: http://dx.doi.org/10.1016/j.foodcont.2013.02.004

[51] Grabowski NT, Klein G. Bacteria encountered in raw insect, spider, scorpion, and centipede taxa including edible species, and their significance from the food hygiene point of view. Trends in Food Science & Technology. 2017;**63**:80-90

[52] Agabou A, Allou i N. Importance of *Alphitobius diaperinus* (Panzer) as a reservoir for pathogenic bacteria in Algerian broiler houses. Veterinary World. 2010;**3**:71-73

[53] Amadi E, Ogbalu O, Barimalaa I, Pius M. Microbiology and nutritional composition of an edible larva (*Bunaea alcinoe* Stoll) of the Niger Delta. Journal of Food Safety. 2005;**25**:193-197

[54] Braide W, Oranusi S, Udegbunam L, Oguoma O, Akobondu C, Nwaoguikpe R. Microbiological quality of an edible caterpillar of an emperor moth. Journal of Ecology and the Natural Environment. 2011;**3**:176-180

[55] Strother K, Steelman C, Gbur E. Reservoir competence of lesser mealworm (Coleoptera: Tenebrionidae) for *Campylobacter jejuni* (Campylobacterales: Campylobacteraceae). Journal of Medical Entomology. 2005;**42**:42-47

[56] Templeton JM, De Jong A, Blackall P, Miflin J. Survival of Campylobacter spp. in darkling beetles (*Alphitobius diaperinus*) and their larvae in Australia. Applied and Environmental Microbiology. 2006;**72**:7909-7911

[57] Wales A, Carrique-Mas J, Rankin M, Bell B, Thind B, Davies R. Review of the carriage of zoonotic bacteria by arthropods, with special reference to Salmonella in mites, flies and litter beetles. Zoonoses Public Health. 2010;**57**:299-314

[58] Chai J, Shin E, Lee S, Rim H. Foodborne intestinal flukes in South-east Asia. The Korean Journal of Parasitology. 2009;**47**:69-102

[59] Simpanya M, Allotey J, Mpuchane S. A mycological investigation of phane, an edible caterpillar of an emperor moth, Imbrasia belina. Journal of Food Protection. 2000;**63**:137-140

[60] Regulation (EU) No. 37/2010 of 22 December 2009 on pharmacologically active substances and their classification regarding maximum residue limits in foodstuffs of animal origin. Official Journal of the European Union. 2010;**20.1.2010**:L15/1

[61] Poma G, Cuykx M, Amato E, Calaprice C, Francois J, Covaci A. Evaluation of hazardous chemicals in edible insects and insect-based food intended for human consumption. Food and Chemical Toxicology [Internet]. 2017;**100**:70-79. Available from: http://dx.doi.org/10.1016/j.fct.2016.12.006

[62] Dillon R, Charnley K. Mutualism between the dessert locust Schistocerca gregaria and its gut microbiota. Research in Microbiology. 2002;**153**:503-509

[63] Jung J, Heo A, Park YW, Kim YJ, Koh H, Park W. Gut microbiota of *Tenebrio molitor* and their response to environmental change. Journal of Microbiology and Biotechnology. 2014;**24**(7):888-897

[64] Stoops J, Crauwels S, Waud M, Claes J, Lievens B, Van Campenhout L. Microbial community assessment of mealworm larvae (*Tenebrio molitor*) and grasshoppers (*Locusta migratoria migratorioides*) sold for human consumption. Food Microbiology. 2016

[65] Garofalo C, Osimani A, Milanovi V, Taccari M, Cardinali F, Aquilanti L, et al. The microbiota of marketed processed edible insects as revealed by high-throughput sequencing. Food Microbiology. 2017;**62**:15-22

[66] Hogenhout S, Ammar D, Whitfield A, Redinbaugh M. Insect vector interactions with persistently transmitted viruses. Annual Review of Phytopathology. 2008;**46**:327-359

[67] Duarte A, Attili-Angelis D, Baron N, Forti L, Pagnocca F. Leaf-cutting ants: An unexpected microenvironment holding human opportunistic black fungi. Antonie Van Leeuwenhoek. 2014;**106**:465-473

[68] Guedes F, Attili-Angelis D, Pagnocca F. Selective isolation of dematiaceous fungi from the workers of *Atta laevigata* (Formicidae: Attini). Folia Microbiology (Praha). 2008;**57**:21-26

[69] Chen X, Feng Y, Chen Z. Common edible insects and their utilization in China: Invited review. Entomology Research. 2009;**39**:299-303

[70] Milanović V, Osimani A, Pasquini M, Aquilanti L, Garofalo C, Taccari M, et al. Getting insight into the prevalence of antibiotic resistance genes in specimens of marketed edible insects. International Journal of Food Microbiology. 2016;**227**:22-28

[71] Post K, Riesner D, Walldorf V, Mehlhorn H. Fly larvae and pupae as vectors for scrapie. Lancet. 1999;**354**:1969-1970

[72] Thackray A, Muhammad F, Zhang C, Denyer M, Spiropoulos J, Crowther D, et al. Prion-induced toxicity in PrP transgenic Drosophila. Experimental and Molecular Pathology. 2012;**92**:194-201

[73] Thackray A, Di Y, Zhang C, Wolf H, Pradl L, Vorberg I, et al. Prion-induced and spontaneous formation of transmissible toxicity in PrP transgenic Drosophila. Biochemical Journal. 2014;**463**:31-40

[74] Thackray A, Zhang C, Arndt T, Bujdoso R. Cytosolic PrP can participate in prion-mediated toxicity. Journal of Virology. 2014;**88**:8129-8138

[75] Nishimune T, Watanabe Y, Okazaki H, Akai H. Thiamin is decomposed due to Anaphe spp. entomophagy in seasonal ataxia patients in Nigeria. The Journal of Nutrition. 2000;**130**:1625-1628

[76] Mathew T, Ndamiso M, Shaba E, Mustapha S, Muhammed S, Adamu A. Phytochemical, physicochemical, anti-nutritional and fatty acids composition of soldier termites (*Coptotermes gestroi*) from Paikoro local government, Niger state, Nigeria. Journal of Environmental Science, Toxicology and Food Technology. 2013;**7**:71-75

[77] Chakravorty J, Ghosh S, Megu K, Jung C, Meyer-rochow VB. Entomology nutritional and anti-nutritional composition of *Oecophylla smaragdina* (Hymenoptera: Formicidae) and *Odontotermes* sp. (Isoptera: Termitidae): Two preferred edible insects of Arunachal Pradesh, India. Journal of Asia-Pacific Entomology [Internet]. 2016;**19**(3):711-720. Available from: http://dx.doi.org/10.1016/j.aspen.2016.07.001

[78] Ji K-M, Zhan Z-K, Chen J-J, Liu Z-G. Anaphylactic shock caused by silkworm pupa consumption in China. Allergy. 2008;**63**:1405-1410

[79] Barre A, Velazquez E, Delplanque A, Bienvenu F, Bienvenu J, Benoist H, et al. Les allergènes croisants des insectes comestibles. Rev française d'allergologie [Internet]. 2016;**56**(7-8):522-532. Available from: http://dx.doi.org/10.1016/j.reval.2016.10.008

[80] Houbraken M, Spranghers T, De Clercq P, Cooreman-algoed M, Couchement T, De Clercq G, et al. Pesticide contamination of *Tenebrio molitor* (Coleoptera: Tenebrionidae) for human consumption. Food Chemistry [Internet]. 2016;**201**:264-269. Available from: http://dx.doi.org/10.1016/j.foodchem.2016.01.097

[81] Gaylor M, Harvey E, Hale R. House crickets can accumulate polybrominated diphenyl ethers (PBDEs) directly from polyurethane foam common in consumer products. Chemosphere. 2012;**86**:500-505

[82] Green K, Broome L, Heinze D, Johnston S. Long distance transport of arsenic by migrating Bogong moths from agricultural lowlands to mountain ecosystems. The Victorian Naturalist 2001;**118:**114-116

[83] Vijver M, Jager T, Posthuma L, Peijnenburg W. Metal uptake from soils and soil–sediment mixtures by larvae of *Tenebrio molitor* (L.) (Coleoptera). Ecotoxicology and Environmental Safety. 2003;**54**:277-289

[84] Handley M, Hall C, Sanford E, Diaz E, Gonzalez-Mendez E, Drace K, et al. Globalization, binational communities, and imported food risks: Results of an outbreak investigation of lead poisoning in Monterey County, California. American Journal of Public Health. 2007;**97**:900-906

[85] Zhuang P, Zou H, Shu W. Biotransfer of heavy metals along a soil-plant-insect-chicken food chain: Field study. Journal of Environmental Sciences. 2009;**21**:849-853

[86] Cappellozza S, Saviane A, Tettamanti G, Squadrin M, Vendramin E, Paolucci P, et al. Identification of *Enterococcus mundtii* as a pathogenic agent involved in the "flacherie" disease in *Bombyx mori* L. larvae reared on artificial diet. Journal of Invertebrate Pathology [Internet]. 2011;**106**(3):386-393. Available from: http://dx.doi.org/10.1016/j.jip.2010.12.007

[87] Inglis D, Sikorowski P. Microbial contamination and insect rearing. In: Schneider J, editor. Principles and Procedures for Rearing High Quality Insects. Mississippi State University; 2009. pp. 325-340

[88] Fraqueza M, Barreto A. HACCP. In: Toldra, Hui Y, Astiasaran I, Sebranek J, Talon R, editors. Handbook of Fermented Meat and Poultry. Oxford: Wiley-Blackwell; 2005. pp. 513-534

Chapter 6

Polyandry as a Signal of Phase Shift in Female Desert Locust *Schistocerca gregaria*

Severin Dushimirimana, David Damien and
Thierry Hance

Additional information is available at the end of the chapter

http://dx.doi.org/10.5772/67938

Abstract

The multiple mating by female (polyandry) is a widespread behavior in insect species. This behavior is known to be a kind of fitness maximization, but some case of sexual selection factors can explain the evolution of this behavior in relation with the phenotype plasticity model. In this paper, we analyze the role of polyandry in the reproductive success and in the phase shift process in the gregarious desert locust. In an applied perspective, knowledge on the reproductive success and in the phase shift process is essential to perform mass rearing for human food production. Our results suggest that multiple mating is not associated with fitness benefits. Polyandry acts as a signal of phase shift through offspring. We showed that hatchlings of gregarious females mated only once are smaller and green at 87.2% in first egg pods and produced the solitary form of the desert locust. The coloration of offspring in females mated with two males reaches only 15.2% of green forms versus 84.8% of mostly blacks. In this study, we showed that females mated more than two times with different males produce larger eggs, heavier, and black hatchlings characteristic of gregarious phase known in *S. gregaria*.

Keywords: polyandry, phase polyphenism, phenotypic plasticity, fitness, *Schistocerca gregaria*

1. Introduction

In males, the number of matings (polygyny) determines the reproductive success and the ability to transmit its own genes to the next generation [1, 2]. Things are less clear in females, as multiple matings (polyandry) represent a loss in time and energy [2]. Several hypotheses have been proposed to explain why natural selection had favored the evolution of polyandry in animals [3, 4]. Polyandrous females may gain direct benefits in fitness maximization, genetic benefits, or reduction of harassment [5]. For example, polyandry stimulates ovulation [6], ensures fertilization, or brings nuptial gifts [2]. Genetic benefits could be paternity choice [6]), sex ratio manipulation particularly in haplodiploid insects [7], inbreeding avoidance by mixed paternity [8], and sperm stored renewal [3].

In *Orthoptera*, for female, multiple mating has many advantages [9]: preventing the females from sperm depletion [10], increasing the rate of oviposition [11–15], or providing material incorporated into eggs or female biomass [16, 17]. Since the 1950, it has been shown that desert locust copulation occurs frequently, and multiple copulation appears to serve mainly as a stimulant to egg-laying [18]. However, until now, adaptive significance of polyandry is poorly understood in *Schistocerca gregaria*. According to our previous results, multiple mating did not seem to be necessary for avoiding sperm depletion [19]. Three other hypotheses can be stated to explain polyandry in this species: (1) an increase of female fitness (fecundity, fertility, and lifespan), (2) an increase in offspring viability as observed in other species [20, 21], and (3) the provision of additional nutrients [22].

At the same time, the mechanisms underlying the phase polyphenism in desert locusts have been studied since many years and various aspects of this phenomenon have been clarified [23–30]. However, multiple matings and male harassment could be also an indication of overcrowding and thus of phase shift. In that context, the role of multiple mating for females in the phase polyphenism determination was never investigated in detail. It is the main objective of this work.

Recently, it has been discovered that progeny characteristics such as egg size, hatchlings body color are predetermined in the ovaries of the mother in *S. gregaria* [31–35]. They underlined a strong evidence of maternal effect controlling egg production and egg size [34, 35].

Green and small hatchings are solitarious locusts, whereas black and heavy hatchings are gregarious individuals [34–39]. In 1932, Faure has already suggested that small hatchlings appear from small eggs produced by solitarious form, whereas large hatchlings are from large eggs produced by gregarious form. Although eggs size alone is inadequate to explain completely the locust polyphenism phenomenon, it remains very important to understand if multiple-mating behavior may influence egg size and thereafter, phase shift.

In this study, we examined the adaptive significance of polyandry of desert locust and more particularly, the impact of multiple mating on phase shift and eggs biomass. In practice, we determined the effects of multiple mating on female fitness components (fecundity, fertility, female longevity, and offspring survival) and offspring phase (egg sizes hatchlings body color and body size).

As this species is usually consumed by human population [40] and may even be a source of dietary or therapeutic sterols [41], a better knowledge of its mating strategies may improve possible mass rearing systems.

http://dx.doi.org/10.5772/67938

2. Materials and methods

Under laboratory conditions, we tested the benefits for a female *S. gregaria* to mate with several males. Five treatments have been analyzed during this survey: a female mated with only one male, or successively with two, three, and four males. We also noted the effect of promiscuity on the fitness traits of a female in presence of four males after mating. As number of egg pods per females varies, we analyzed and compared only the two first egg pods that were laid.

2.1. Insect rearing

S. gregaria individuals were provided by the Laboratory of Entomology of the Zoological Institute of KUL (Leuven, Belgium). These individuals descend from locusts that were originally collected in Nigeria [42]. Rearing of gregarious *S. gregaria* was conducted in cages (50 × 60 × 80 cm) containing 100–200 individuals per cage. Rearing temperatures were 35 ± 2°C (12 h day) and 28 ± 2°C (12 h night) with a relative humidity of 60%. Locusts were fed on fresh cabbage leafs and rolled oats. Females and mature imago were collected in cages few days before sexual maturation and separated by sex to ensure virgin status.

2.2. Mating test and fitness analyses

Mature females were placed individually with mature males in experimental cages for mating. Pairs were observed until the male disengaged his genitalia and males were then removed to prevent re-mating. To ensure that mated females were inseminated, only females that copulated for more than 4 h according to our previous results [19] were used. Among these females, single-mated (n = 23) have been considered as control. Plastic cups filled with clean, moist sand were placed in cages to collect egg pods. Egg pods were incubated at 30°C, and cages were checked daily for the hatched larvae. After hatching, the oothecae were removed from the sand the number of nonhatched eggs was counted.

The total number of first-stage larvae plus nonhatched eggs represents the female fecundity and the proportion of hatched larvae and its fertility.

The once-mated remaining females were marked with a red permanent marker (Artiline®-70N) for the re-mating tests. Marked mated females were placed with a virgin male 24 h after the first mating. After the second copulation, females (n = 26) were allowed to lay eggs until death. Third (n = 21) and fourth (n = 27) re-mating events were conducted in the same way. In the last treatment, females (n = 24) were allowed to mate once and stay in contact with a group of four males until female death. Fecundity and fertility were examined as explained above.

Polyandrous females, used in re-mating tests, were monitored daily to measure their lifespan. Survival of offspring produced by multiple or a one-time-mated female was also determined until imaginal moult. Offsprings were maintained in the same conditions of rearing as explained above, and dead individuals were counted.

2.3. Effects of multiple mating on phase transformation

To characterize the phase transformation, we considered three parameters: egg size, hatchling body color, and body weight.

2.3.1. Eggs collection and eggs weight measurements

Three treatments were used: females mated with single males, females mated with two males, and females mated with three males.

After copulation, females were removed from copulation cage and held individually in small cages.

Plastic cups filled with clean, moist sand were placed in cages to collect egg pods. Cages were observed, and eggs were carefully collected from egg pods after 6–12 h. Eggs were washed with distilled water to remove ovipositing substrate or sand and then separated individually and put in marked Petri dishes (9 cm in diameter; 1.5 cm in height), according to the number of males that female has been mated. Eggs were then weighed individually using a Taledo balance at 0.1 mg.

2.3.2. Hatchling body color and body weight

Five treatments were used: females mated with single male, females mated with two males, females mated with three males, females mated with four males, and females allowed to mate once and stay in contact with a group of four males until female death. Mating tests and eggs incubation until hatchling were conducted as explained above. The body color of hatchlings was observed 6 h after hatching. Nymphs from single mating and multiple mating were divided into two hatchling color groups: green and black including dark spots to entirely black. Hatchlings from single and multiple matings were weighted 6 h after hatching using Ohaus Explorer balance (0.01 mg) to analyze body weight.

2.4. Statistical tests

Repeated measures analysis of variance (ANOVA) with multiple matings as a factor was performed on fitness data characteristics: fecundity, fertility, and hatchling body weight. Univariate analysis of variance was used to compare egg size produced by females mated with single male or more than one male. For offspring survival, the log-rank test was used to compare survival curves with GraphPad Prism. In all experiments, an assumption of normality of residuals was tested and statistic significance was 0.05.

3. Results

3.1. Fecundity

The fecundity did not differ between the treatment in first egg pods (one-way ANOVA $F_{4,116} = 0.37$ $P = 0.82$), with a mean ranging from 60.9 ± 10 ($n = 23$) eggs in single-mated to 60.6 ± 8.3

($n = 26$), 59.2 ± 6.8 ($n = 21$), 59 ± 8.1 ($n = 27$), and 61.3 ± 8 ($n = 24$) eggs in multiple-mated treatment. Fecundity did not differ also in second egg pods (one-way ANOVA $F_{4,65} = 0.32$, $P = 0.85$) with a mean ranging from 58.9 ± 7.2 ($n = 11$) eggs in single-mated to 57.3 ± 7 ($n = 13$), 57.7 ± 6.8 ($n = 14$), 58.7 ± 9 ($n = 15$), and 56.2 ± 5.5 ($n = 16$) eggs in multiple-mated treatment.

3.2. Fertility

In two first egg pods, we found no significant difference among treatment. Fertility in first eggs pods did not differ between the treatment (one-way ANOVA $F_{4,116} = 0.49$ $P = 0.73$). For the first egg pods, a mean (±SD) of 55.4 ± 9.5 larvae were observed in single-mated ($n = 23$) and 55.8 ± 7.2 ($n = 26$), 53.4 ± 6 ($n = 21$), 54.1 ± 6.7 ($n = 27$), and 55.8 ± 7.3 ($n = 24$) larvae, respectively, in multiple-mated treatments; indicating the hatchling success of 90.9 ± 5% in single-mated, 92.3 ± 4.1% in two males mated, 90.4 ± 5.5% in three males mated, 91.6 ± 4.7% in four males mated, and 91.1 ± 5.6% in male promiscuity mating.

No statistical differences appeared among treatments in second egg pods (one-way ANOVA $F_{4,64} = 0.24$ $P = 0.91$). The hatchling success of 90.7 ± 7.5% in single-mated, 92.3 ± 5.1% in two males mated, 92.7 ± 6.3% in three males mated, 91.2 ± 5.4% in four males mated, and 92 ± 5.1% in male promiscuity.

3.3. Female longevity

The female life span after copulation was 37.1 ± 21.9 days (mean ± SD) in single-mated and 45 ± 18.6, 38.1 ± 18.5, 44.3 ± 23.1 days, respectively, in two, three, and four males mated treatments. The life span was 44.7 ± 23.8 if female mated stay with a group of four males.

There was no significant difference between the five treatments (one-way ANOVA $F_{4,116} = 0.76$, $P = 0.54$).

3.4. Offspring survival

With a log-rank test, no difference was found among offspring whose mother has been mated once or multiple times (Chi-square: 2.372 *df*: 5, $P > 0.05$). Survival rate at imaginal moult was 77% in single-mated offspring ($n = 44$), 73% in two males mated offspring ($n = 52$), 80% in three males mated offspring ($n = 50$), 79% in four males mated offspring ($n = 62$), 74% in one male and one female pair offspring ($n = 54$), and 85% in one female paired with four males offspring ($n = 60$).

3.5. Phase transformation

3.5.1. *Eggs weight*

The number of matings had a highly significant effect on the eggs weight (likelihood ratio test: 14.82, $df = 2$, $P = 0.0006$). The females mated once have an estimated mean egg weight of 6.69 mg. Our results showed a significant relationship between eggs weights and the number of males mated with the females ($a = 0.47$ $t = 32.17$, $P < 0.0001$) (**Figure 1**).

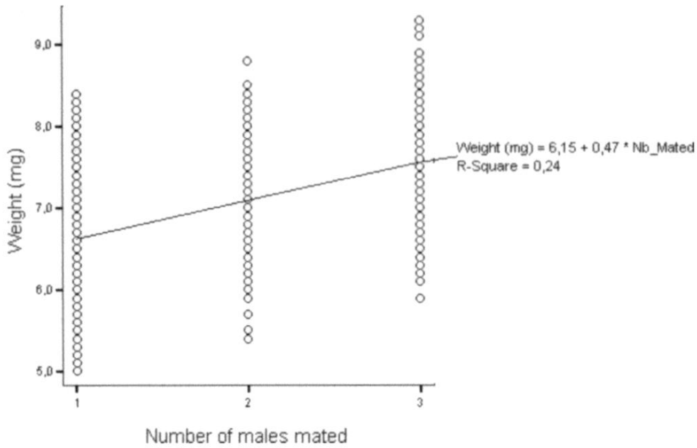

Figure 1. Relationship between number of males mated by the female and eggs weight.

3.5.2. *Nymphs weight*

Our results showed that hatchlings from single male-mated females are smaller than the hatchlings from the females mated with more than two males (Student-Newman-Keuls multiple comparisons test $P < 0.0001$). Hatchlings weight was 12.4 ± 1.3 mg, $n = 80$ (mean ± SD) in single-mated, 16.1 ± 1.8 mg, $n = 78$ in two males mated, 16 ± 1.2 mg, $n = 74$ in three males mated, and 15.9 ± 1.2 mg, $n = 77$ in four males mated treatment. Hatchlings weight of female mated and stay with four males was 16.2 ± 1.5 mg, $n = 76$.

3.5.3. *Hatchling coloration*

Hatchlings from single-mated female were green at 87.2% ($n = 23$) in first egg pods (**Figure 2**) and 85.5 ($n = 11$) in second egg pods (**Figure 3**). The coloration change in females mated with two males was only 15.2% green versus 84.8% black in first egg pods ($n = 26$) (**Figure 2**), 13.6% green

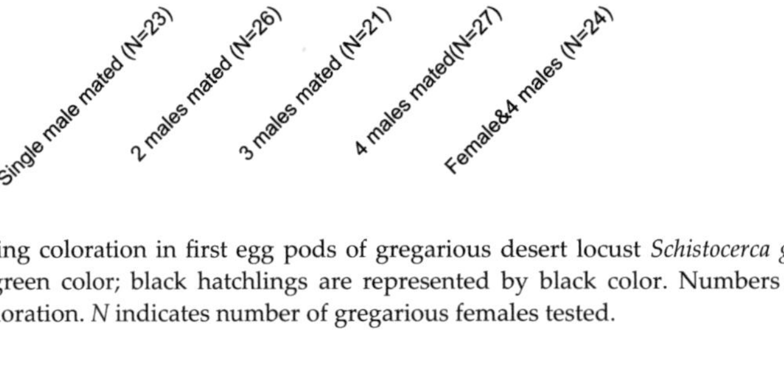

Figure 2. Hatchling coloration in first egg pods of gregarious desert locust *Schistocerca gregaria*: green hatchlings are represented as green color; black hatchlings are represented by black color. Numbers inside columns indicate the percentage of coloration. *N* indicates number of gregarious females tested.

http://dx.doi.org/10.5772/67938

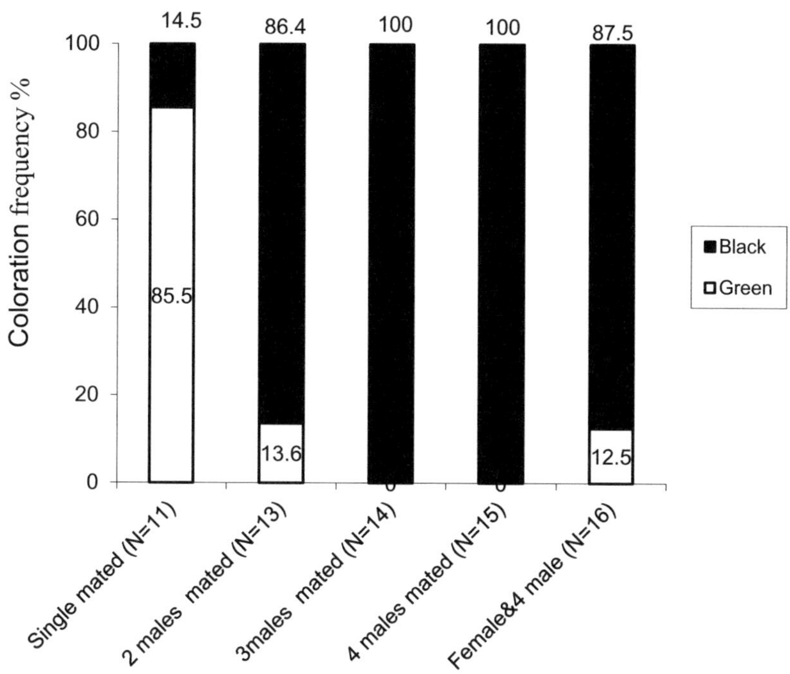

Figure 3. Hatchling coloration in second egg pods of gregarious desert locust *S. gregaria*: green hatchlings are represented as green color; black hatchlings are represented by black color. Numbers on columns indicate the percentage of coloration. *N* indicates number of gregarious females tested.

larvae versus 86.4% black larvae in second egg pods ($n = 13$) (**Figure 3**). Hatchlings from females mated with three and four males were 100% black in first and second egg pods (**Figures 2** and **3**).

4. Discussion

In this paper, we experimentally varied the number of copulations with different males to elucidate the significance of polyandry in gregarious desert locust *S. gregaria*. *S. gregaria* is a very important agronomic pest but presents also many possibilities in terms of food production with an average of 76% of protein content [40]. Our most important finding is that desert locust polyphenism may be induced by multiple matings that act as a signal of phase shift. We show that when a female is mated with two or more males, she produced a majority of black hatchlings with a higher biomass, which correspond to the gregarious phase. On the contrary, females mated only once produce smaller hatchlings with the green coloration of the solitary phase.

Many factors have been proposed to explain the parental effect by which desert locust control their progeny phase: (1) rearing density [18, 43]; (2) pheromonal factor derived from the accessory gland of the female parent [36–39, 44]; (3) ovary influence [32]. Multiple mating can be considered as a reliable signal of overcrowdings leading on a biparental influence on egg size and thus on phase transformation. The desert locust polyphenism is a case of phenotypic plasticity with a numerous convergent evolutionary events [27, 35, 37, 45]. Recent investigations on insect polyphenism have concluded that the expression of a polyphenism is governed by one or more hormones secretion [46]. Thus fluctuations in the concentration of hormones can be a regulator of polyphenism expression. This concentration can be modulated by multiple mates in three main processes in desert locust: (1) males ejaculate signal; (2) males courtship and mating femur stimulating; and (3) male-derived nutrients. In the first hypothesis, males introduce sperm with a hormonal factor, which could have a gregarizing effect at some concentration level [4, 32, 37] or a hormonal factor that acts on the ovarioles to control egg size [35]. In the second hypothesis, female femur is stimulated during male courtship and mating [4, 32]; a stimulus is transmitted to the locust brain and causes a hormonal factor controlling egg size [35, 47]. Both processes are not mutually exclusive. Finally, male-derived nutrients during multiple mates could influence quantitatively egg size, polyandrous female receiving more nutrients, and produce larger eggs producing the gregarious forms. More experimental studies must be made to confirm these hypotheses. However, the egg weight significantly increases in the progeny of females mated more than once but irrespectively of the number of matings. This indicates that it is the fact to mate at least twice that induces a change in egg size and progeny color more than the number of matings. A potential nutritive contribution of each male ejaculate is probably thus not the main factor.

Polyandry is known to increase fitness [1, 48–51], and numerous cases were reported in *Orthoptera* [12, 15, 52–54]. However, except egg size, no fitness benefits could be underlined. Indeed, none of the life-history traits investigated (i.e., female fecundity, fertility, lifespan, and offspring survival) were improved by multiple mating. This could be linked to the fact that the number of eggs produced is less important than their size and the possibility it confers to hatchlings to migrate under stressful conditions. These results may help to improve mass rearing of this species as a source of food or nutrient.

In conclusion, we showed another form of phenotypic plasticity of desert locust in which the expression of numerous physiological, morphological, and behavioral traits occurs in response to multiple matings behavior.

Acknowledgements

We thank Burundian Government for scholarship to realize this study in Belgium. Thanks also to Dr Frederic Muratori for helpful comments on the manuscript and establishment of desert locust rearing at Earth and Life Institute (UCL). We thank the Laboratory of Entomology at Zoological Institute of Leuven (Belgium), particularly Prof. Huybrechts for providing the insects to start the rearing.

Author details

Severin Dushimirimana[1,2]*, David Damien[1] and Thierry Hance[1]

*Address all correspondence to: dusev2002001@yahoo.fr

1 Biodiversity Research Center, Earth & Life Institute, Université Catholique de Louvain, Louvain-la-Neuve, Belgium

2 Department of Natural Sciences, Ecole Normale Supérieure du Burundi, Burundi

References

[1] Arnquist G. and Nilsson T.: The evolution of polyandry: Multiple mating and female fitness in insects. Animal Behaviour. 2000; 60: 145-164.

[2] Thornhill R. and Alcock J.: The evolution of insect mating systems. 1983, London: Harvard University Press.

[3] Halliday T. and Arnold S.J.: Multiple mating by females: A perspective from quantitative genetics. Animal Behaviour. 1987; 35: 939-941.

[4] Reynolds D.J.: Animal breeding systems. Trends in Ecology and Evolution. 1996; 11: 68-72.

[5] Stockley: Sexual conflict resulting from adaptations to sperm competition. Trends in Ecology and Evolution. 1997; 12: 154-159.

[6] Birkhead T.R. and Møller A.P.: Sperm competition and sexual selection. 1998, 24-28, London, UK. Academic Press; 826 p.

[7] Godfray H.C.J.: Parasitoids: Behavioral and evolutionary ecology. 1994. USA: Princeton University Press.

[8] Brooker M.G., Lowley I., Adams M. and Baverstock P.R.: Promiscuity: An inbreeding avoidance mechanism in a socially monogamous species?. Behavioral Ecology and Sociobiology. 1990; 26: 243-258.

[9] Parker G.A., Smith J.L.: Sperm competition and the evolution of the precopulatory passive phase behavior in *Locusta migratoria migratorioides*. Journal of Entomology Series A. General Entomology. 1975; 49: 155-171.

[10] Reinhardt K. and Köhler J.: Costs and benefits of mating in the grasshopper *Chorthippus parallelus* (*Caelifera: Acrididae*). Journal of Insect Behavior. 1999; 12(3): 283-293.

[11] Friedel T. and Gillott C.: Contribution of male-produce proteins to vitellogenesis in *Melanplus sanguipes*. Journal of Insect Physiology. 1977; 23: 145-151.

[12] Lange A.B., Loughton B.G.: An oviposition-stimulating factor in the male accessory reproductive gland of the locust, *Locusta migratoria*. General and Comparative Endocrinology. 1985; 57: 208-215.

[13] Lafon-Cazal M., Gallois D., Lehouelleur J., Bockaert J.: Stimulatory effects of male accessory-gland extracts on the myogenicity and the adenylate cyclase activity of the oviduct of *Locusta migratoria*. Journal of Insect Physiology. 1987; 33: 909-915.

[14] Ridley M.: Mating frequency and fecundity in insect. Biological Reviews. 19881; **1**(63): 509-549.

[15] Walker P.M., Lewis C.J., Whitman D.W.: Effects of males on the fecundity and fertility of female *Romalea microptera* grasshoppers. Journal of Orthoptera Research. 1999; 8: 277-283.

[16] Hinn J.C. and Feaver M.N.: Spermatophore number transferred increases linearly with copulation duration in *Melanoplus differentialis* (*Orthoptera: Acrididae*). Journal of Orthoptera Research. 2001; 10(2): 147-152.

[17] Christian C.V., Kretzschmar A.S., Speakman J.R. and Lehmann G.U.C.: Female bush-crickets fuel their metabolism with male nuptial gifts. Biology Letter. 2008; 4: 476-478.

[18] Hunter-Jones P.: Laboratory studies on the inheritance of phase characters in locusts. Anti-locust Bulletin. 1958; 29: 1-32.

[19] Dushimirimana S., Damiens D., Hance T.: Comparison of reproduction traits of regular and irradiated male desert locust *Schistocerca gregaria* (*Orthoptera*: *Acrididae*): Evidence of last-male sperm precedence. Biology Open. 2012; 1: 1-5.

[20] Teng Z.-Q. and Kang L.: Egg-hatching benefits gained by polyandrous female locusts are not due to the fertilization advantage of nonsibling males. Evolution. 2007; 61(2): 470-476.

[21] Tseng H.-F., Yang R.-L., Lin C., Horng S.-B.: The function of multiple mating in oviposition and egg maturation in the seed beetle *Callosobruchus maculatus*. Physiological Entomology. 2007; 32: 150-156.

[22] Olsson M., Gullberg A., Tegelström H., Madsen T. and Shine R.: Can female adders multiply?. Nature. 1994; 369: 528.

[23] Faure J.C.: The phases of locusts in South Africa. Bulletin of Entomological Research. 1932; 23: 293-405.

[24] Uvarov B.P.: Grasshoppers and locusts. A handbook of general acridology. 1966. Vol. 1, London: Cambridge University Press.

[25] Uvarov B.: Grasshoppers and locusts. 1977; Vol. 2. London, UK: Centre for Overseas Pest Research.

[26] Dale J.F., Tobe S.S.: The endocrine basis of locust phase polymorphism. In: Chapman, R.F., Joern, A. (Eds.), Biology of grasshoppers. 1990, New York: John Wiley and Sons: 393-414.

[27] Pener M.P.: Locust phase polymorphism and its endocrine relations. Advances in Insect Physiology. 1991; 23: 1-79.

[28] Pener M.P., Yerushalmi Y.: The physiology of locust phase polymorphism: An update. Journal of Insect Physiology. 1998; 44: 365-377.

[29] Pener M.P. and Simpson S.J.: Locust phase polyphenism: An update. Advances in Insect Physiology. 2009; 36: 1-272.

[30] Verlinden H., Badisco L., Marchal E., Wielendaele V.P., Jozef Vanden Broeck J.V.: Endocrinology of reproduction and phase transition in locusts. General and Comparative Endocrinology. 2009; 162: 79-92.

[31] Tanaka S., Maeno K.: Phase-related body-color polyphenism in hatchlings of the desert locust, *Schistocerca gregaria*: Re-examination of the maternal and crowding effects. Journal of Insect Physiology. 2006; 52: 1054-1061.

[32] Tanaka S., Maeno K.: Maternal effects on progeny body size and color in the desert locust, *Schistocerca gregaria*: Examination of a current view. Journal of Insect Physiology. 2008; 54: 612-618.

[33] Maeno K., Tanaka S.: Maternal effects on progeny size, number and body color in the desert locust, *Schistocerca gregaria*: Density-and reproductive cycledependent variation. Journal of Insect Physiology. 2008b; 54: 1072-1080.

[34] Maeno K., Tanaka S.: Artificial miniaturization causes eggs laid by crowdreared (gregarious) desert locusts to produce green (solitarious) offspring in the desert locust. *Schistocerca gregaria*. Journal of Insect Physiology. 2009; 25: 849-854.

[35] Tanaka S., Maeno K.: A review of maternal and embryonic control of phase-dependent progeny characteristics in the desert locust. Journal of Insect Physiology. 2010; 56: 911-918.

[36] McCaffery A.R., Simpson S.J., Islam M.S., Roessingh P.: A gregarizing factor present in the egg pod foam of the desert locust *Schistocerca gregaria*. Journal of Experimental Biology. 1998; 201: 347-363.

[37] Simpson S.J., McCaffery A.R., Hägele B.F.: A behavioural analysis of phase change in the desert locust. Biological Reviews of the Cambridge Philosophical Society. 1999; 74: 461-480.

[38] Hägele B.F., Oag V., Bouaïchi A., McCaffery A.R., Simpson, S.J.: The role of female accessory glands in maternal inheritance of phase in the desert locust *Schistocerca gregaria*. Journal of Insect Physiology. 2000; 46: 275-280.

[39] Simpson S.J., Miller G.A.: Maternal effects on phase characteristics in the desert locust, *Schistocerca gregaria*: A review of current understanding. Journal of Insect Physiology. 2007; 53: 869-876.

[40] Zielińska E., Baraniak B., Karaś M., Rybczyńska K. and Jakubczyk A.: Selected species of edible insects as a source of nutrient composition. Food Research International. 2015; 77: 460-466.

[41] Cheseto X., Kuate S.P., Tchouassi D.P., Ndung'u M., Teal P.E. and Torto B.: Potential of the desert locust *Schistocerca gregaria* (*Orthoptera*: *Acrididae*) as an unconventional source of dietary and therapeutic sterols. PloS One. 2015; 10(5): e0127171.

[42] Hoste B., Luyten L., Claeys I., Clynen E., Rahman M.M., DeLoof A. and Breuer M.: An improved breeding method for solitarious locusts. Entomologia experimentalis et applicata. 2002; 104: 281-288.

[43] Islam S.M., Roessingh P., Simpson J.S., A.R. Mccaffery: Effects of population density experienced by parents during mating and oviposition on the phase of hatchling desert locusts. *Schistocerca gregaria*, Proceedings: Biological Sciences. 1994; 257: 93-98.

[44] Simpson S.J., Sword A.G., De Loof: Advances, controversies and consensus in locust phase polyphenism research. Journal of Orthoptera Research. 2005; 14: 213-222.

[45] Sword G.A., Lecoq M., Simpson S.J.: Phase polyphenism and preventative locust management. Journal of Insect Physiology. 2010; 56: 949-957.

[46] Nijhout H.F.: Insect hormones. 41 William Street, Princeton, New Jersey 08540: Princeton University Press. 1994: p. 260.

[47] Highnan K.C., Haskel P.T.: The endocrine systems of isolated and crowded Locusta and Schistocerca in relation to oocyte growth, and the effect of flying upon maturation. Journal of Insect Physiology. 1964; 10: 849-864.

[48] Simmons L.W.: The contribution of multiple mating and spermatophore consumption to the lifetime reproductive success of female field crickets (*Gryllus bimaculatus*). Ecological Entomology. 1988; 13: 57-69.

[49] Madsen T., Shine R., Loman J. and Häkansson T.: Why do female adders copulate so frequently?. Nature. 1992; 355: 440-441.

[50] Keller and Reever: Why do females mate with multiple males? The sexually selected sperm hypothesis. Advances in the Study of Behaviour. 1995; 24: 291-315.

[51] Zeh J.A. and Zeh D.W.: The evolution of polyandry II: post-copulatory defenses against genetic incompatibility. Proceedings of Royal Society London B. 1997; 264: 69-75.

[52] Gwynne: Courtship feeding increases female reproductive success in bushcrickets. Nature. 1984; 307: 361-363.

[53] Butlin R.K., Woodhatch C.W., Hewitt G.M.: Male spermatophore investment increases female fecundity in a grasshopper. Evolution. 1987; 41: 221-225.

[54] Tregenza T, Wedell N: Benefits of multiple mates in the cricket *Gryllus bimaculatus*. Evolution. 1998; 52: 1726-1730.

Chapter 7

Nutritive Evaluation of Earthworms as Human Food

Zhejun Sun and Hao Jiang

Additional information is available at the end of the chapter

http://dx.doi.org/10.5772/intechopen.70271

Abstract

Earthworms have been a traditional medicine in China for at least 2300 years. Because of medicine food homology in China, people have been using earthworms as a food for several centuries. Earthworms are rich in protein and various amino acids; the protein content of earthworm meal was 54.6–59.4% on dry weight. Their protein content and amino acid composition are better than those of fish meal, cow milk and soybean meal. The crude fat content of earthworm meal was 7.34%. Earthworm protein is easily hydrolyzed into free amino acids. The hydrolyzed body fluids contain 9.34% protein and 78.73 mg of free amino acids per liter of raw fluid and are rich in vitamins and minerals. By the fast development of molecular biological techniques, more Verm pharmaceuticals and functional components were isolated from earthworms. An antibacterial peptide and a functional earthworm powder were introduced in this paper. In short, based on its nutrient content and functional components, earthworms could be an excellent raw material source as homology of medicine and food for human use, especially as functional food in the future.

Keywords: earthworm, protein, nutrition, functional food and active peptide

1. Introduction

It is a time-honored, traditional custom to use live earthworm as angling bait, fish feed and duck feed. However, only within last 30 years has earthworm culture spread throughout the world as an excellent protein source replacing fish meal and soybean meal on a commercial scale since the potential for use of some earthworms in accelerating waste decomposition and protein production on a large scale was reported. It has been a well-known fact that earthworms can be used in the stabilization of organic wastes. Not only in the laboratory but also in practice has it been demonstrated that *Eisenia fetida*, the species most intensively studied in this respect, grows and reproduces well in many kinds of waste, converting them into live protein-rich feed and dark odorless casts with good physical structure for organic fertilizer.

Locally in China, people have been using earthworms as a food for several centuries. In ancient times, people in Fujian and Guangdong provinces of China had the habit of eating earthworms. Even now, in Taiwan and Henan and Guangdong provinces, some local people prepare special dishes featuring the earthworm as a basic ingredient. Records from Chinese ancient book "On Guo Yi Gong" say that the people who lived in Fujian, who were considered different from other people, considered earthworms to be a delicacy. They cut the earthworms in small pieces and mixed them with meat filling to make their food tastier. Even now, earthworm soup, a traditional delicacy, is still offered in some restaurants of Guangdong province [1].

Earthworms are also consumed by Ye'kuana Amerindians of the Alto Orinoco of Venezuela (Paoletti *et al.*, 2002, private communication). In recent years, some countries in Western Europe and the Southern East have produced various earthworm products such as canned earthworms, mushroom-earthworms and earthworm biscuits and bread. In California, a company composed of several earthworm farms in Northern America held an exhibition and competition on earthworm food in 1975. In some African and South American countries, earthworms are commonly eaten. Owing to the high content and good quality of earthworm protein, the high content of vitamin B and other bioactive substances, it is very likely that earthworms could become an important source of animal protein in human nutrition if other sources become limited.

As early as in the 1940s, Lawrence and Miller reported the nutrient composition of earthworms [2]. Now functional molecules have been refined [3, 4]. But these results showed considerable variability due to variations in these wild earthworm species and little attention was paid to earthworms as a potential protein source. Since the 1970s, some people have evaluated the earthworms as animal feeds; it has been well known that the earthworm contains elevated levels of protein and other nutritive compounds and has a high potential value as livestock feed.

However, the most common uses of earthworms are to treat some diseases in traditional Chinese medical practices. "Earth dragon" in Chinese is the name given to the earthworm or "white neck earthworm," recorded early in an ancient medicinal book "Shen-nong-ben-cao-jing." Another famous ancient medicinal book, "Ben-cao-gang-mo," listed 42 entries relating to earthworms in the 42 volumes about insects and described their medicinal use in detail. Now "earth dragons" used as medicine are mainly two species: *Pheretima aspergillum* and *Allolobophora caliginosa trapezoides*. However, besides the two species of earthworm above, the following species are also used as Chinese medical animal materia: *Pheretima carnosa, P. medioca, P. hupeiensis, P. posthuma, P. praepinguis, P. tschiliensis, P. tschiliensis lanzhouensis, P. guilleimi, P. vulgaris, P. peetinifera* and *E. fetida* [5]. Besides the medicinal use of whole earthworms, some pharmaceutical ingredients and special active proteins or peptides have been studied in recent years [6].

In the middle of 1980s, a Japanese scientist extracted an enzyme from earthworms that can dissolve thrombi in experimental conditions. This enzyme preparation has been made into an oral medicine by the pharmaceutical industry for use in the prevention of cardiovascular disease in China, Japan and Korea [7]. With the development of modern biochemical science in recent years, great progress has been achieved in the field of research on isolating some active compounds, including small molecular proteins, peptides and amino acids [8].

The section of proteins and nutritive components of earthworm most cited from the paperof Sun et al. [9] on earthworm as a potential protein resource, which was published in not socommonly available journal Ecology of Food and Nutrition (see [9]).

2. Proteins and nutritive components in earthworms

We selected a common cultivated species *E. fetida*, commercial name "Daping II," as a standard earthworm for nutritive evaluation [9]. Chemical analysis procedures of earthworm were according to standard methods [10]. T tests and Duncan's new multiple range test, cited from Steel and Torrie [11], were applied to separate means when treatment effects were significant. Data in **Tables 1–5** represent mean standard deviation (SD). The results of nutrient composition of the earthworms are summarized in **Table 1**, which also shows some common animal food and feed ingredients. A comparison of the amino acid contents of earthworm meal and casts with that of fishmeal, eggs, cow milk, wheat bran and two kinds of insect is given in **Tables 2** and **3**.

Feed or food	DM	CP	Fat	Ash	Ca	P	ME (kcal/g)
Fresh earthworm (*E. fetida*)	15.7 (±1.47)	11.02 (±0.46)	1.89 (±0.50)	1.4 (±0.01)	0.22 (±0.10)	0.65	
Earthworm meal (*E. fetida*)	90.6 (±2.56)	54.6 (±0.92)	7.34 (±0.60)	21.2 (±0.05)	1.55 (±0.12)	2.75	2.99*
Earthworm casts (*E. fetida*)	82.2 (±1.58)	7.9 (±0.26)	1.1 (±0.43)	34.2 (±0.07)	1.42 (±0.05)	0.28	0.95*
Peruvian fish meal	90.8 (±0.72)	62.0 (±0.28)	9.7 (±0.48)	14.4 (±0.13)	3.91 (±0.04)	2.90	2.90
Chinese fish meal	88.5 (±1.81)	53.9 (±0.09)	9.3 (±0.48)	18.9 (±0.12)	4.59 (±0.09)	2.15	2.35
Cow milk	12.7 (±0.43)	3.5 (±0.29)	3.5 (±0.11)	0.7 (±0.08)	0.12 (±0.04)	0.09	0.65*
Egg	26.3 (±0.34)	12.9 (±0.17)	11.5 (±0.08)	1.0 (±0.01)	0.05 (±0.08)	0.21	1.63**
Soybean meal	88.1 (±1.28)	43 (±0.70)	5.4 (±0.13)	5.9 (±0.08)	0.32 (±0.11)	0.50	2.64
Corn meal	86.5 (±0.49)	8.6 (±0.31)	3.5 (±0.09)	1.4 (±0.02)	0.04 (±0.08)	0.21	3.32
Wheat bran	82.2 (±1.62)	14.2 (±0.10)	2.0 (±0.33)	4.4 (±0.04)	0.14 (±0.21)	1.06	1.78

Data are presented as % (weight/weight on dry matter basis) [9]. DM, dry matter; CP, crude protein; Ca, calcium; P, phosphorus and ME, metabolic energy.

*Calculated figure [12].

**Food energy.

Table 1. Nutrient composition of earthworm (***Eisenia fetida***) and some feed and foods.

Amino acid	Earthworm meal	Earthworm casts	Peruvian fish meal	Chinese fish meal	Hen egg	Raw cow milk	Wheat bran
Thr*	2.72 (±0.09)	0.46 (±0.01)	2.88 (±0.28)	2.22 (±0.24)	2.42 (±0.06)	1.20 (±0.08)	0.45 (±0.038)
Ser	2.71 (±0.08)	0.46 (±0.09)	2.63 (±0.14)	2.01 (±0.24)	3.64 (±0.14)	1.57 (±0.09)	0.74 (±0.036)
Gly	3.12 (±0.24)	0.49 (±0.03)	4.26 (±0.09)	3.26 (±0.25)	1.58 (±0.32)	0.54 (±0.11)	0.84 (±0.033)
Cys	0.42 (±0.10)	0.09 (±0.03)	0.56 (±0.18)	0.42 (±0.22)	1.16 (±0.20)	0.22 (±0.09)	0.33 (±0.028)
Val*	2.39 (±0.27)	0.44 (±0.01)	2.80 (±0.25)	2.29 (±0.37)	3.26 (±0.18)	1.57 (±0.14)	0.67 (±0.014)
Met*	1.01 (±0.42)	0.19 (±0.03)	1.65 (±0.57)	1.64 (±0.33)	1.6 (±0.09)	0.68 (±0.16)	0.15 (±0.003)
Ile*	2.40 (±0.12)	0.38 (±0.026)	2.42 (±0.48)	2.23 (±0.40)	2.99 (±0.011)	1.28 (±0.18)	0.37 (±0.042)
Leu*	3.94 (±0.15)	0.78 (±0.034)	4.28 (±0.22)	3.85 (±0.19)	4.20 (±0.10)	2.58 (±0.15)	0.80 (±0.042)
Tyr	1.73 (±0.08)	0.24 (±0.42)	2.12 (±0.26)	1.63 (±0.12)	1.98 (±0.24)	1.28 (±0.017)	0.52 (±0.042)
Phe*	2.12 (±0.81)	0.31 (0.038)	2.68 (±0.28)	2.10 (±0.45)	2.73 (±0.25)	1.46 (±0.19)	0.48 (±0.034)
Lys*	4.26 (±0.50)	0.68 (±0.033)	4.35 (±0.34)	3.64 (±0.27)	3.32 (±0.22)	2.11 (±0.20)	0.47 (±0.017)
His*	1.36 (±0.24)	0.12 (±0.014)	1.66 (±0.21)	0.90 (±0.32)	1.16 (±0.23)	0.72 (±0.25)	0.35 (±0.002)
Arg	3.27 (±0.33)	0.64 (±0.17)	3.87 (±0.37)	3.02 (±0.51)	2.90 (±0.13)	0.89 (±0.12)	0.95 (±0.027)

Amino acid abbreviations used in **Tables 2–9**: Ala, alanine; Arg, arginine; Asp, aspartic acid; Cys, cysteine; Gln, glutamine; Glu, glutamic acid; Gly, glycine; His, histidine; Ile, isoleucine; Leu, leucine; Lys, lysine; Met, methionine; Phe, phenylalanine; Pro, proline; Ser, serine; Thr, threonine; Trp, tryptophan; Tyr, tyrosine and Val, valine.

*Essential amino acids for humans.

Table 2. Contents of amino acids in earthworm and common feeds and foods [9] (weight/weight on a dry matter basis).

Amino acid	*Kuru* body (n=2)	*Kuru* gut organs[a] (n=2)	*Motto* body (n=2)	*Motto* smoked (n=2)
Asp	71.3	35.5	62.5	68.1
Thr	34.2	23.2	30.1	32.4
Ser	35.8	18.6	32.2	34.8
Glu	124	66.2	107	109
Pro	26.2	15.9	23.2	23.1
Gly	49.4	28.2	39.1	34.2
Ala	42.9	26.6	37.1	36.9
Val	33.9	21.4	31.5	32.5
Met	17.0	11.2	14.4	16.0
Ile	33.8	25.3	29.6	30.5
Leu	62.0	34.9	55.3	55.4
Tyr	21.4	9.52	20.2	19.9
Phe	28.9	18.3	26.7	27.3
His	18.7	11.1	15.8	14.7
Lys	54.2	32.1	49.7	48.5
Arg	60.5	32.9	55.7	53.5

Amino acid	*Kuru* body (n=2)	*Kuru* gut organs[a] (n=2)	*Motto* body (n=2)	*Motto* smoked (n=2)
Cys	7.17	5.02	5.81	5.86
Trp	8.23	5.54	8.51	9.64
Total protein content	729	421	644	653

a:Parts not eaten.

Source: Paoletti *et al.* [13].

Table 3. Amino acid content of ***motto*** and ***kuru*** (mg·g^{-1} dry weight) [9] (the values reported are the means of two determinations on two different samples).

	E. fetida wild	*E. fetida* cultured	*A. caliginosa*	*P. guillemi*
Protein	59.11±2.96 (39.9±2.56)	59.41±2.23 (44.8±1.49)	49.7±1.71 (31.9±2.95)	50.11±1.36 (31.22±2.37)
Lys	4.17 (3.22)	4.25 (3.77)	3.04 (2.05)	2.86 (1.65)
Met	1.13 (0.81)	0.99 (0.95)	0.82 (0.50)	0.75 (0.40)
Ala	5.22 (2.94)	5.39 (2.75)	3.42 (1.94)	3.10 (1.83)
Arg	4.06 (2.95)	3.89 (3.26)	1.42 (1.02)	3.02 (1.74)
Cys	0.73 (0.47)	0.65 (0.35)	0.50 (0.30)	0.61 (0.44)
Val	2.89 (2.09)	3.29 (2.30)	2.27 (1.25)	2.08 (1.19)
Phe	2.38 (1.60)	2.46 (1.91)	1.70 (1.11)	1.52 (0.89)
Thr	3.40 (2.08)	2.94 (2.66)	2.18 (1.48)	2.01 (1.50)
His	1.56 (0.85)	1.74 (1.22)	0.62 (0.40)	1.04 (0.62)
Ile	2.36 (1.83)	2.60 (2.10)	2.03 (1.54)	1.91 (1.52)

Bracket data in parentheses refer to earthworm meal with gut inclusions.

Table 4. Contents (% weight/weight on dry matter basis) of protein and amino acids in meal of different earthworm species [9].

FAA	Body fluids (g/l)	Fresh bodies (g/kg)	FAA	Body fluids (g/l)	Fresh bodies (g/kg)
Asp	5.9±0.09	9.2±0.28	Ile	4.7±0.11	4.2±0.38
Thr	4.3±0.08	4.7±0.14	Leu	9.4±0.24	5.3±0.36
Ser	3.1±0.33	4.4±0.08	Tyr	3.2±0.22	4.1±0.32
Glu	10.4±0.15	3.3±0.22	Phe	4.2±0.08	4.9±0.12
Gly	2.7±0.42	4.2±0.19	Lys	7.2±0.17	7.2±0.21
Ala	6.0±0.12	9.0±0.26	His	1.8±0.23	2.9±0.15
Cys	/	0.6±0.09	Arg	5.5±0.14	8.0±0.09
Val	5.4±0.10	6.1±0.08	Prs	2.1±0.24	3.3±0.18
Met	3.0±0.09	2.9±0.17			

Table 5. Contents of free amino acids (FAA) in body fluids and fresh body of *E. fetida* [9].

Since Lawrence and Miller first reported the protein content of earthworm in 1945, many nutritional evaluations of the earthworm have been published [14–21]. From the literature, it can be concluded that earthworms contain high levels of protein and that this protein is rich in the amino acids, considered essential for food of domesticated animals and humans. However, the protein content varies with the earthworm species and experimental food of earthworms, and the protein contents reported range from 48 to 71% (dry weight basis). *E. fetida* is a species with relatively high protein content of 58–71% dry weight or about 9.7% of its live weight [1, 17]. Generally speaking, on dry weight basis, earthworm protein contents was from 48% to 71%, the species of *E. fetida* was 58-71%. On live weight basis, earthworm serous fluid contained only about 9.7% of crude protein. Our results show that the protein content of *E. fetida* was *ca* 55% dry weight, which is similar to the results of Fosgate and Babb [22], but lower than those reported in Sabine's review [23].

Table 4 summarizes the contents of protein and amino acids of four earthworm species, including three wild earthworm species of *Allolobophora caliginosa, Pheretima guillemi* and *E. fetida* besides cultivated *E. fetida*. A comparison of protein and amino acids contents of four earthworm species showed that larger wild species *A. caliginosa* and *P. guillemi* was significantly lower than smaller species *E. fetida* in protein content; but there was no difference of the same species *E. fetida* in between wild and cultivated. It was found that different protein content among earthworm species was perhaps due to variations in gut inclusions. As mentioned above, why large earthworm species, *A. caliginosa* and *P. guillemi,* present lower protein content? It may result from the incomplete excretion of gut contents. If taking longer time for larger species to excrete the gut contents, there may be no significant differences in protein contents between the various species.

The chemical analysis results of earthworm serous fluid, which made of *E. fetida* by adding proteinase was to hydrolyze into a fluid [24], showed in **Table 5**. By this method, 78.73 g/l of free amino acid (FAA) was gotten, and more than 90% of the earthworm protein was hydrolyzed into FAA. The hydrolyzed also contained rich minerals, trace elements and vitamins (**Table 6**).

Few reports are available on the protein content of earthworm fluids because earthworms are usually applied as livestock feed in the form of dry earthworm meal. Zou [25] reported that earthworm body fluids contained a range of 7–9% protein.

Element	mg/l	Element	g/l	Vitamin	mg/l
Mn	3.27±1.01	Ca	0.33±0.046	A	13.46±0.37
Zn	6.90±0.27	Na	0.31±0.038	B_1	54.65±0.80
Cu	1.08±0.30	K	0.99±0.042	B_2	83.06±1.05
Pb	0.30±0.15	Fe	0.33±0.072	E	31.64±0.64
Se	0.30±0.12	Mg	0.11±0.025	C	292.00±2.35

Table 6. Contents of some elements and vitamins in body fluids of ***E. fetida*** [9].

The results from the literature show that the contents of essential elements, such as copper (Cu), manganese (Mn) and zinc (Zn), are between one and six times higher in earthworm meal than in soybean meal and fish meal [1, 25, 26]. A similar trend was found in the present study regarding the mineral element contents of earthworm body fluids, which contained three to six times more of most elements than did fish meal and soybean meal (**Table 5**). Earthworms seem able to tolerate large concentrations of some metals in their tissues, which may be related to specific metal-binding proteins in earthworms [27]. These metal-binding proteins may affect nutrient absorption when earthworms are introduced into feeds and foods; for example, a 90% phosphorus utilization rate from an earthworm-based diet was reported [1].

3. Pharmaceutical and other composition of earthworms

Earthworms contain lumbrofebrine, terrestrolumbrlysin, lumbritin, hypoxanthine and other purines, pyrimidines, choine and guanidine. The fat of earthworm is composed of octade acids, palmitic acids, high-chain unsaturated fatty acids, linear and odd carbon fatty acids, branched fatty acids, phosphatide, cholesterin, etc. The yellow chloragenous cells and organs of *Lumbricus terrestris* contain rich amounts of carbohydrates, lipids, protein, pigments and some alkaline amino acids. The yellow pigment perhaps consists of riboflavin or its analogues [28].

The tissues of *Pheretima* spp. contain large amounts of microelements, Zn 59.1 μg/g, Cu 25.4 μg/g, Fe 1735.5 μg/g, Cr 10.93 μg/g, Mo 0.25 μg/g, Ca 1019.2 μg/g and Mn 1143 μg/g [29]. These of Allolobophora *caliginosa* contain crude protein 57.96%; crude fat 6.53%, crude ash 21.09%, crude fiber 0.36%, and N extract 14.06%. These of Eisenia *fetida* contain crude protein 64.61%; crude fat 12.29%, crude ash10.16%, crude fiber 0.27%, and N extract 12.67% [24].

The blood and body fluids of *L. terrestris* contain small concentrations of glucose (0.01–0.05 μg/ml) [30], considerably lipids, including 35.14% neutral fat, 41.74% glucolipid and 23.12% phosphatide. The C-chain of the fatty acid is between 10 and 22°C. The neutral fat consists mainly of laurel acid, oleate, myristic acid and decanoic acid. The fatty acids of glucolipids are decanoic acid and some short-chain fatty acids. The fatty acids of phosphatide are oleate, decanoic acid, linoleate and behenic acids. The proportion of unsaturated fatty acids is higher than that of neutral fatty acids and saccharides [31]. A P-peptide substance exists in the gut wall of *L. terrestris* [32].

Some active enzymes occur in the yellow chloragenous cells and organs of *L. terrestris* in high concentrations, including catalase, peroxidase, dismutase, β-D-glucosyl enzyme, alkaline phosphatase, esterase, S-amino-γ-ketoglutaric dehydrogenase and porphyrin synthetase. The body fluids of *Eisenia* spp. contained at least 18 proteins with molecular weights between 1000 and 95,000 Da [33].

Kaloustain (1986) reported that earthworm species of *A. caliginosa* contain a protein, which can hydrolyze collagen. Scientists from Japan, China and Korea found and separated enzymes from the earthworm gut and body fluids, which can dissolve fibrin. These enzymes have been developed as innovative medicines to treat cerebral thromboses and myocardial infarction. Wang *et al.* [34] found and separated a kind of acid antibacterial peptide, tetra decapeptide, which has produced a disease-resistant, nutrient earthworm preparation, that can be used in plant and animal production. Six antimicrobial peptides were isolated and purified from earthworm tissue liquid homogenate and coelomic fluid, which contained 5–50 amino acid (AA) residues with the same or similar sequence of Ala-Met-Val-Ser-Gly, and named the antibacterial Verm peptides family (AVPF) according to their structure and antibacterial characteristics [34].

3.1. Earthworm antibacterial properties and polypeptide characterization

Earthworm's active protease and functional peptides have wide development prospect as substitution for fortune green medicine and food. An antibacterial peptide is described as following:

A_{3-4}, a antibacterial peptide from earthworm, was carried out by digestive test with pronase E treatment at 37°C for 4h. After polypeptide was disintegrated, and yellow precipitation in the solution and showed antibacterial activities (**Table 7**).

3.2. Spectrum of earthworm antibacterial peptides

Table 8 showed that the diameter of inhibitory ring of *Erwinia carotovora* is the biggest and inhibitory ring of *Xanthomonas campestris* as well as *E. coli* are second; that of *Staphylococcus aureus's* is the smallest. But A has strong inhibitory effect on *S. aureus*. With the enhancing purity from A, A_3, to A_{3-4-2}, the bacteriostatic activity on various bacteria is depressed except on *X. campestris*. This shows that earthworm antibacterial peptides have wide antibacterial chart, and there are differences among various bacterial strains in their antibacterial efficiency. A has obvious inhibitory effect on *Agrobacterium tumefaciens*, which can cause the formation of plant tumor tissue, but A_3 and A_{3-4-2} have no effect on this bacterium.

3.3. Earthworm antibacterial peptide A_{3-4-2} amino acid composition

After A_{3-4-2} was hydrolyzed in HCl and an analysis showed that the peptide is made of 10 kinds of amino acids, of which Glu included Gln (**Table 9**). The results of alkaline hydrolysis showed that it contained tryptophane. The result suggested that the A_{3-4-2} is composed of 40 amino acid residues belonging to 11 kinds of amino acids

Treatment	The diameter of inhibitory ring (mm)
With pronase E	1.5
Normal saline	0
Without pronase E	6

Table 7. Effect of pronase E on bacteriostatic activity of the A_{3-4}.

The diameter of inhibitory ring (mm)	*E. coli*	*Staphylococcus aureus*	*Xanthomonas campestris*	*Erwinia carotovora*	*Agrobacterium tumefaciens*
Crude extract (A)	4.4	4.3	4.1	4.9	3.8
A_3	4.3	3.5	4.5	3.3	–
A_{3-4-2}	4.32	3.21	4.43	3.4	–

Table 8. Bacteriostatic activity of worm antibiotics on various bacteria.

Residue kind	Reside numbers	Residue kind	Reside numbers
Asp	4	Ile	/
Thr	1	Leu	4
Ser	1	Tyr	/
Gln	2		
Glu	6	Phe	/
Gly	10	Lys	2
Ala	8	His	/
Cys	1	Trp	1
Val	/	Arg	/
Met	/	Pro	/

Table 9. Amino acid composition of earthworm antibacterial peptide A_{3-4-2}.

4. The actual nutritional value of earthworm protein

The nutritional value of protein depends upon its specific amino acid composition. Comparing the amino acid contents of *E. fetida* reported in our study with those in other reports [1, 15, 23, 24] and with those in *Eudrilus eugeniae. Lumbricus rubellus* [35, 36], *A. caliginosa* and *P. guillemi* [1], there is considerable variability in amino acid contents among the species and even within the same species. Nevertheless, some research suggested that the contents of individual amino acids differ between species by no more than 17% and usually by considerably less [16].

Sabine [23] reviewed the amino acid compositions of earthworm protein reported by five authors, and compared these data with that of two common sources of protein supplements, meat meal and fish meal. Sabine's results showed that valine, leucine and isoleucine were higher in earthworm meal than in fishmeal, but lower than in meat meal. The methionine content of earthworm meal was close to that of meat meal, but 200% of that in fish meal. Arginine, histidine and phenylalanine contents in earthworm meal were close to those in meat meal and slightly higher than in fishmeal. Threonine, cystine and tryptophan in earthworm meal were significantly higher than in fishmeal and meat meal. Therefore, Sabine suggested that since earthworm protein was high in essential amino acids, including the sulfur-containing amino acids, it should be very suitable for animal feed [23].

In our studies, the contents of EAA (on a dry matter basis) in earthworm meal were close to or slightly lower than those in fishmeal and eggs. The methionine content of our earthworm meal was significantly less than that of fish meal, soybean meal and hen egg ($p < 0.05$), which contrasts with data in Sabine's review [23]. The lysine content of our earthworm meal was significantly higher than that of eggs and cow milk ($p < 0.05$).

From the literature cited and the results from our present study, it can be concluded that *E. fetida* (especially when cultured) is relatively high in most essential amino acids, compared to those in other common foods and feeds. Biological value and net protein utilization are the two most important parameters used conventionally to evaluate protein quality of feed materials. Some researchers reported a biological value of 84% and a net protein utilization of 79% in a rat growth assay with *E. fetida* protein. These results were verified in fish and chicken tissues [37–39].

Earthworm protein is easily dissolved by enzymes into free amino acids. This suggests that earthworm protein is easily metabolized by animals. Thus, earthworms seem a promising source of protein supplementation not only for animal feed but also for human food.

5. An earthworm protein powder as a functional food

Modern scientific research indicated earthworms also contain some unknown nutrients, which have important physiological activities or functions for human health and body remediation. In December 2012, the Chinese government listed earthworm protein as a new resource of food.

Recently, a new product of earthworm protein powder has been developed. People use the earthworms as the main material, through hydrolysis extraction, ultrafiltration, nanofiltration and spray drying process, then they got the earthworm protein powder. In the powder, there are lots of water-solubility small molecular proteins.

The earthworm protein powder is a light-yellow powder and there are no visible impurities in it. It tastes sweet with a special flavor. The total protein content according to the provisions of the GB/T5009.5-2003 detection method, the minimum protein content is greater than or equal to 75%. The product standard of the earthworm protein powder is given in **Table 10**.

Materials	**Content**
Total amino acid, g/100g	≥75
Thr, g/100g	≥2.5
Val, g/100g	≥5.0
Met, g/100g	≥0.8
Phe, g/100g	≥4.0
Ile, g/100g	≥1.0
Leu, g/100g	≥8.0
Lys, g/100g	≥7.0

Materials	Content
Fe, mg/kg	≥200
Zn, mg/kg	≥20
Ca, mg/kg	≥2.0
Total numbers of colony, cfu/g	≤1000
Escherichia coli, cfu/g	≤40
Mould, cfu/g	≤25
Microzyme, cfu/g	≤25
Salmonella spp.	0
Shigella spp.	0
Staphylococcus aureus	0
Hemolytic streptococcus	0

Table 10. Product standard of the earthworm protein powder.

6. The safety of earthworms as a protein source

Due to the assimilation of metals from their environment, earthworms often contain elevated levels of metals, including some heavy metals that could be harmful to animals and human beings [40, 41]. However, in experiments with fish and chicken, no significant increases in heavy metals was found in the carcasses of the trial animals fed on earthworms. Earthworm consumption may pose other possible hazards. For instance, it has been suggested that in natural food chains earthworms might carry some parasitic nematodes of chicken and pigs [1]. However, earthworms have not been found to be significant in the natural distribution of parasitic nematodes in chickens or in pigs [42]. Increasing proportions of earthworms and earthworm casts in fish and chicken diets did not produce any significant changes of the organoleptic qualities of the meat produced, although the earthworm meal itself is reported to have a garlic taste. Continuous monitoring of the organic substrates used in vermiculture, as well as of earthworm composition and contamination, will be necessary to safeguard human health, especially where large-scale commercial earthworm production is considered for human consumption.

7. Conclusions

1. Earthworm (*E. fetida*) meal has a high protein content of 54.6–59.4% of its dry weight. Its protein content and amino acid composition are close to those of Peru fishmeal and eggs and better than those of Chinese fish meal, cow milk and soybean meal. The crude fat content of earthworm meal was 7.34%. The ash and energy content of earthworm meal are the highest of all the materials tested except for the higher energy content of corn meal.

2. Earthworm protein is easily hydrolyzed into free amino acids. Earthworm body fluids contain 9.34% protein and 78.73 mg of free amino acids per liter of raw fluid and are rich in vitamins and minerals, particularly iron (Fe).

3. The protein content of earthworms shows considerable variability between different species and between different experimental treatments in the same species, possibly due to variability in the extent of gut inclusion.

4. Based on its nutrient content, earthworm could be an excellent raw material source as homology of medicine and food for human use, especially as functional food. A new functional food of earthworm protein powder made of earthworm has been developed in China.

Acknowledgements

The section of proteins and nutritive components of earthworm most cited from the paper of earthworm as a potential protein resource, which was published in Ecology of Food and Nutrition (see [9]).

Author details

Zhejun Sun and Hao Jiang*

*Address all correspondence to: jianghao66066@sina.com

Beijing Vermitech Institute, China Agricultural University, Beijing, China

References

[1] Zeng ZP, et al. Earthworm Cultures. Wuhan, China: Hubei People's Publishing House; 1982. pp. 1-146

[2] Lawrence RD, Miller HR. Protein content of earthworms. Nature 1945;155-157

[3] Bianka S, Kedracka-Krok S, Panz T, Morgan AJ, Falniowski A, Grzmil P, Plytycz B. Lysenin family proteins in earthworm coelomocytes. Comparative Approach, Developmental & Comparative Immunology. 2017;**67**:404-412

[4] Li C, Chen M, Li X, Yang M, Wang Y. Purification and function of two analgesic and anti-inflammatory peptides from coelomic fluid of the earthworm, *Eisenia foetida*. Peptides. 2017;**89**:71-81

[5] Gao. Records of Chinese Materia Medica. Jilin Sci. & Tech. Press; 1996. pp. 18-27 (in Chinese)

[6] Li W, Wang C, Sun Z. Vermipharmaceuticals and active proteins isolated from earthworms. Pedobiologia. 2011;**54**(Suppl):S49-S56

[7] Chen H, et al. Earthworm fibrinolytic enzyme: Anti-tumor activity on human hepatoma cells in vitro and in vivo. Chin. Med. J. (Engl.) 2007;**120**(10):898-904

[8] Sun Z. Earthworm as a biopharmaceutical: from traditional to precise. European Journal of BioMedical Research. 2015;**1**(2):28-35

[9] Sun Z J, Liu X C, Sun L H, et al. Earthworm as a potential protein resource. Ecology of Food and Nutrition. 1997;**36**(2-4):221-236

[10] Xu ZY, et al. Experimental Handbook for Nutrition Analysis. Publishing House for Agriculture; 1979. pp. 1-98

[11] Steel RGD, Torrie JH. Principles and Procedure of Statistics. New York, NY: McGraw-Hill; 1960

[12] Xiao ZD. A survey of earthworm metabolic energy. Journal of Jilin Agricultural University. 1984;**4**:28-32

[13] Paoletti MG, Buscardo E, Vanderjagt DJ, et al. Nutrient content of earthworms consumed by Ye'Kuana Amerindians of the Alto Orinoco of Venezuela. Proceedings Biological Sciences 2003. 2002;**270**(1512):249-257

[14] Istiqomah L, Sofyan A, Damayanti E, et al. Amino acid profile of earthworm and earthworm meal (*Lumbricus rubellus*) for animal feedstuff. Journal of the Indonesian Tropical Animal Agriculture. 2009;**34**(4):253-257

[15] Mclnroy DM. Evaluation of the earthworm *Eisenia fetida* as food for man and domestic animals. Feedstuffs. 1971;**43**:37-46

[16] Avnish C. Vermitechnology, Vermiculture, Vermicompost and Earthworms. LAP LAM BERT Academic Publishing; 2012

[17] Serna-Cock L, Rengifo-Guerrero CA, Rojas-Restrepo MA. Use of earthworm (*Eisenia foetida*) flour and hydrolyzed chicken feathers as sources of nitrogen and minerals for ethanol production. Waste & Biomass Valorization. 2017;**1**:1-10

[18] Blair B. Unusually nutrition a look at some feed alternatives. Milling. 1985;**168**(6):21-22

[19] Boushy E. Biological conversion of poultry and animal waste to a feedstuff for poultry. Poultry Adviser. 1986;**19**(6):51-65

[20] Edwards CA. Production of feed protein from animal waste by earthworms. Philosophical Transactions of the Royal Society, B: Biological Sciences 1985;**310**:299-307

[21] Edwards CA, Neuhauser EF, editors. Earthworms in Waste and Environmental Management. The Hague, The Netherlands: SPB Academic; 1988

[22] Fosgate OT, Babb MR. Biodegradation of animal waste by *Lumbricus terrestris*. Journal of Dairy Science 1972;**55**:870-872

[23] Sabine JR. Earthworms as a source of food and drugs. Earthworm Ecology. Netherlands: Springer; 1983. pp. 285-296

[24] Zhang HZ. Nutritional composition and evaluation of earthworms. Chinese Journal of Zoology. 1984;**19**(2):18-19

[25] Zou WC. Nutritive composition of earthworm's body fluids and its exploitation and utilization. Chinese Journal of Zoology. 1993;**28**(3):7-11

[26] Beyer WN. Relation of pH and other soil variables to concentration of Pb, Cu, Zn, Cd and Se in earthworms. Pedobiologia 1987;**30**(3):167-172

[27] Suzuki KT, Yamamura M, Mori T. Cadmium-binding proteins induced in the earthworm. Archives of Environmental Contamination & Toxicology. 1980;**9**(4):415-424

[28] Jiangsu New Medical College. Dictionary of Materia Medica. Shanghai Sci. & Tech. Press; 1985. p. 2111

[29] Zhao C. Element analyses of some earthworms. Bulletin of Chinese Materia Medica. 1988;**13**(9):38

[30] Mitsuhiro U, Yamashita T, Nakazawa M, et al. Purification and characterization of a new serine protease (EF-SP2) with anti-plant viral activity from *Eisenia foetida*: Analysis of anti-plant viral activity of EF-SP2. Process Biochemistry. 2011;**46**(9):1711-1716

[31] Lee H, Han. Guk yongyang siklyang hakhoechi. 1986;**15**(2):115-118

[32] Cho JH, et al. Lumbricin I, a novel proline-rich antimicrobial peptide from the earthworm: purification, cDNA cloning and molecular characterization. Biochimica et Biophysica Acta 1998;**1408**(1):67-76

[33] Cheng GF, et al. Some active enzymes in earthworms. Journal of Biochemistry 1985; **1**(5/6):161-168

[34] Wang C, Sun Z, Liu Y, Zhang X, Xu G. A novel antimicrobial vermipeptide family from earthworm *Eisenia fetida*. European Journal of Soil Biology 2007;**43**:S127-S134

[35] Taboga L. The nutritional value of earthworms for chickens. British Poultry Science 1980;**21**:405-410

[36] Nadazdin M, Viduc D, Jakobcic Z, et al. Essential Amino Acid Values in Dehydrated Body Weight of Californian Worms with Possible Use in Poultry Nutrition. Veterinarski Glasnik; 1988

[37] Koh TS. Biological utilization of earthworm (*Eisenia fetida*) cake protein supplemented with methionine, histidine or tryptophan in chicks. Korean Journal of Animal Sciences. 1985;**27**(5):305-309

[38] Velasquez L. A note on the nutritional evaluation of earthworm meal (*Eisenia fetida*) in diets for rainbow trout. Animal Production. 1991;**53**(1):119-122

[39] Chakrabarty D, Das SK, Das MK, Biswas P, Karmegam N. Application of vermitechnology in aquaculture. Dynamic Soil Dynamic Plant. 2009;**7**:41-44

[40] Štolfa I, Velki M, Vuković R, Ečimović S, Katanić Z. Effect of different forms of selenium on the plant-soil-earthworm system. Journal of Plant Nutrition & Soil Science. 2017;**180**(2):231-240

[41] Fischer E, Koszorus L. Sublethal effects, accumulation capacities and elimination rate of As, Hg and Se in the manure earthworm, *Eisenia fetida* (Oligochaeta, Lumbricidae). Pedobiologia 1992;**36**(3):172-178

[42] Jakovijevic DD. Some aspects of the epizootology and economical significance of ascariasis in swine. Acta Veterinaria-Beograd 1975;**25**:315-325

Chapter 8

Intelligent and Smart Packaging

Marcello Nicoletti and Paola Del Serrone

Additional information is available at the end of the chapter

http://dx.doi.org/10.5772/intechopen.68773

Abstract

Urgent need of increased food production and availability are crucial in humankind future. A key aspect concerns the food preservation, wherein packaging is a main aspect. Packaging is still in a primitive form, utilized as a way to separate food from environmental conditions, not considering the inside situation. Improvement of packaging means less cost, more food available, and waste decrease. Several solutions are emerging to face this challenge. They are focused on three aspects: the antimicrobial agent; the packaging material; and the technological implication in the final production of packaging. Biotechnology is expected to play a central role in the future food to solve central points, as retain integrity and actively prevent food spoilage. In this phase, several projects are moving, still waiting to converge in adequate products. The galaxy of smart packaging is rapidly moving and increasing in researches. This phase represents a chaotic period of several proposals production and tentative in solving the food preservative problems, using new technologies and advanced techniques, like nanotechnology. Waiting the collapse into a central paradigm, it is interesting and useful to follow the scenario of researches on smart packaging based on natural products here reported.

Keywords: packaging, food waste, food shelf life, food-borne pathogens, food preservation

1. Introduction

'Feed the Planet. Energy for life' was the theme of World Expo 2015 in Milan city that tackled the great problem of sustainable progress and production of future foods. The event projected the feeding as main challenge for humankind and showed the extreme urgency of elements of innovation in technology and science connected to the field of food, in order to contrast feed problems that still today plague several areas of the world. Food production is growing rapidly, as a result of the increasing demand. The global meat production and consumption are supposed to increase from 233 million tonnes (2000) to 300 million tonnes (2020), and milk from 568 to 700 million tonnes over the same period. Egg production should also increase by 30% [1]. The challenge is the possibility of production of enough food for the incoming seven billions of human inhabitants of the planet.

This forecast shows in particular a massive increase in animal protein demand, needed to satisfy the growth in the human population, wherein billions of people ask for an increase of caloric input and better food. The considerable and increasing demand for animal protein is focusing attention on the sources of feed protein and their suitability, quality, and safety for future supply. In addition the quantitative production aspect, there will be a need for considerable increase in feed manufacture, requiring a thriving, successful and modern feed industry, including a key aspect concerning the protection and preservation of the produced and marketed food.

This aspect is strictly related to the safety issues, which will remain paramount in the mind of consumers following recent food crises. Continuing investment is needed in quality assurance programs to gain market access for animal products and to retain consumer confidence [2].

Biotechnology is expected to play a central role in the future food, in order to solve central points, as to retain integrity and actively prevent food spoilage (shelf life). Nowadays, packaging is still mainly a primitive form to separate food from environmental conditions, not considering the inside situation. Increase of the shelf life means reduction of cost and waste. Nowadays, simple material made of paper or plastic are used for packaging. These materials are main part of waste in industrial country and the cost for recycling is increasing as well as the damages to the marine environmental. 'Smart packaging' is focusing the interest in possible solutions. Smartness packaging covers a number of functional methods that can be tailored depending on the product being packaged, including several types of food, beverages, pharmaceuticals, household products, etc. Examples of current and future functional 'smartness' would be in packages that can not only actively prevent food spoilage (to add shelf life) but can also face other aspects in order to reduce the food waste, and eventually maintain, and enhance product attributes (e.g. look, taste, flavor, aroma, etc.), responding actively to changes in product or package environment, confirming product authenticity, and acting to counter theft.

2. Smartness and packaging

The galaxy of smart packaging is rapidly moving and increasing in research and proposals. We can consider this phase as a period of production of several proposals and tentative in solving the food preservative problems, using new technologies and advanced techniques recently available, like Nanotechnology and Molecular Biology. Waiting the collapse into a central paradigm, it is interesting and useful to follow the scenario of research on the subject and the lines of different emerging products.

The novel sustainable solutions in packaging must consider the need for ensuring the safety and quality of food and reducing losses and minimize the environmental impact. As a matter of fact, food packaging plays a crucial role not only in preserving food during distribution and storage from farm to fork but it also contributes to the generation of waste. The aim of modern food packaging systems is focused on the potential capacity to extend the shelf life of perishable foods, by reducing the need for additives and preservatives, and at the same time considering changes in quality. Several methods and approaches, like oxygen scavenging and antimicrobial technologies associated to the production of modified films, are actually considered.

A broad classification of packaging of food can be comprised into four types: Passive, Active, Intelligent, and Smart packaging [3]. They are different solutions to serve the basic and fundamental properties of package: protect, preserve, and present. So far, the dominant packaging is the basic one, based on low-cost material and no interaction with the inside food. In this passive packaging, the traditional packaging systems are included, as the use of covering material, characterized by some inherent insulating, protective, or ease of handling qualities. Usually, the ordinary packaging is not able to preserve the food and is a source of a great quantity of waste with enormous damage to the environment. This situation is increasing in consideration of the growing of consumers in emerging countries, where these consequences are not adequately considered. The packaging is considered active, when the package can interact in same way and/or react to various stimuli, in order to keep the internal environment favorable for the maintenance of the quality of the products. Several environmental, biotic and abiotic factors must be considered, in order to face positively the degradation process. The involved activity could be the presence of oxygen scavenger (an oxygen scavenger can absorb high-energy oxygen inside a package and therefore increase the shelf life of product) or anti-ROS (scavenger of radicals by oxygen or other origins) activity. Smart packaging relies on the use of chemicals, electrical, electronic or mechanical technology or any combinations of them. In particular, smart packaging involves the use of technology that adds feature such that packaging becomes an irreplaceable part of the whole product. According to what above reported, interest in the use of active and intelligent packaging systems for agricultural fresh products has increased in recent years. Active packaging refers to the incorporation of additives into packaging systems, with the aim of maintaining or extending fresh vegetable or livestock products quality and shelf life, while intelligent packaging systems are those that monitor the condition of packaged foods to give information regarding the quality of the packaged food during transport and storage [4].

Besides, the development of intelligent packaging system through the use of sensor technologies indicators (including integrity, freshness, and time-temperature indicators (TTI)) and radio frequency identification (RFID) has been evaluated for potential use in meat and meat products.

The active and smart packaging performs additional functions to the basic one and can be supported by intelligent packaging solutions. Intelligent packaging refers to the introduction of innovations in the design of packaging, with conveniences for the user and usefulness for the consumer or firms in the supply chain. In this way, the product can respond to stimuli generated by the environment or from the product being packaged. It reflects the change in a manner that makes the product more available, more useful, and more long life [5–7].

3. Reduce food waste by retaining integrity and preventing food spoilage

A large quantity of food is lost or wasted throughout the supply chain, from initial agricultural production down to final household consumption. It has been estimated that as much as half of all food grown is lost or wasted before and after it reaches the consumer [8]. That refers to the high perishable food as fresh fruit and vegetables (FFVs) and livestock products. Approximately one-third of all FFVs produced worldwide are lost along food supply chain (FSC) production [9]. High loss rates are associated with a lack of packing houses in India, with FFVs generally packed in the field and some even transported without transit packaging [10]. Food waste occurs at different points in the FSC. Market evolution in reducing waste has enormous potential to develop FSC infrastructure and reduce waste in developing and BRIC (this grouping acronym refers to the countries of Brazil, Russia, India, and China), considering some differences. Significant losses occur even early in the food supply chains in the industrialized regions and the last step of selling. On the other side, in low-income countries, food is lost mostly during the early and middle stages of the food supply chain due to deterioration; much less food is wasted at the consumer level. It is important to consider that many countries of Africa were totally auto-sufficient in food productions and limited but efficient distribution. Nowadays, food is not anymore sufficient in several parts of Africa, generating massive migrations of humans toward Europe as never reported in history records. Among the possible causes there is wrong utilization of technology's opportunities.

4. The microorganisms attack

The quality of food is a complex argument. Nowadays food is the final step of a complex series of events, changing radically the nature of the starting material. The ordinary consumers cannot acquire the quality of most of the food available, but it is the most important element of food. Most of the available food cannot be consumed directly after the production and therefore needs a form of maintenance. During the packaging, the storage, and the shelf life, food is subjected to the attack of microorganisms. These microorganisms are programmed to demolish progressively the molecular structure of the food, as soon and as completely as possible. It is only a matter of time, every structure once living is subjected to be destroyed and converted to be reutilized into a new molecular vivant structure. Food is only the intermediate step between the different forms of life. In other words, the good food is in completion with its recycling, and, working on this limbo, we can be able to efficiently utilize the food, when it is still available. The intermediate step must help to be more efficient than the molecular demolition generated by bacteria. Therefore, in preserving the food, we are working in a thermodynamically unfavorable situation. This usually means cost for low temperature or for disinfection or other methods to retard the bacterial attack. To the biotic factors, we must add the abiotic factors such as humidity and oxygen present in the air. The preserving solution mainly consists of the use of plastic packaging. Once again the solution is not sustainable and terrible in the long-term consequences. As a matter of fact, plastic is covering our planet and causing immense damages to any type of environment. On the other side, low temperature and other forced environmental conditions means a relevant need of energy and consequent considerable cost.

The traditional concept of packaging, in sense of material projected to protect food from physical, chemical, and biological risks, can be overcome. The modified atmosphere or vacuum has led to the diffusion of 'active packaging', systems capable of interacting dynamically with the food and/or with the atmosphere in order to save the healthiness of the product and extend its shelf life. New attempts of solutions are starting from the concept of intelligent and interactive packaging. Once again, nowadays the cost of these types of packaging are not competitive, but the next future will ask for these solutions as necessary and indispensable [11].

5. The after production and food maintaining

Therefore, loss of food derived by contamination of food and spoilage by microorganisms is a major concern to food industries, involving directly consumers and indirectly government authorities. The reduction of these losses is recognized as an important component of improved food security [12]. In developing and emerging economies, this would require market-led large-scale investment in agricultural infrastructure, technological skills and knowledge, storage, transport, and distribution in order to reduce food waste in terms of food losses (decrease in food quantity) and of food spoilage (decrease in food quality), as well of food safety (spread of food-borne diseases).

The technologies used to increase the storage time and to ensure the safe consumption of highly perishable products, such as meat, have undergone a continuous evolution over the time [13–15]. Besides, the diffusion of active packaging, systems capable of interacting dynamically with the food and/or with the atmosphere in order to save the healthiness of the product and to extend its shelf life, is increasing [16]. The need to use materials, more sustainable and more compatible with food, creates a new market and leads to an intense research of natural substances usable for the production of biodegradable wrapping and edible coatings.

The effectiveness of these systems has been improved by the use of film activated by antibacterial substances and chemical or natural preservatives slow release. In recent years, interest in the production and use of active, as well as intelligent packaging systems for meat and meat products, has increased [17, 18]. Also the incorporation of natural antimicrobial substances into edible films has attracted great interest as alternative to control or reducing the growth of food-borne and spoilage microorganisms [19]. Higher consumer acceptance and the economically viable packaging systems are necessary to realize of these packaging technologies. The effectiveness of these systems was improved with the use of antibacterial substances to activate films using chemical or natural preservatives slow release. The shelf life and the safety of foods, especially those processed, are increased by antimicrobial packaging (AP) active against spoilage microorganisms and/or pathogens [20, 21].

The technologies used to increase the storage time and to ensure the safe consumption of highly perishable products are in continuous evolution over the time, in response to the needs of consumers and industry, and availability of new solutions based on advancement in scientific knowledge and technological advancements.

It needs to consider some limitations and cautions in the use of antimicrobials for meat preservation including inactivation of compounds on contact with the meat surface or dispersion of compounds from the surface into the meat mass. There are several mechanisms concerning this aspect, including incorporation of bactericidal compounds into meat products, resulting into their partial alteration by muscle components known to affect significantly the efficacy of the antimicrobial substances and their release. Therefore, physicochemical characteristics of muscle could alter the activity of antimicrobials. Furthermore, water activity of the meat could influence the antimicrobial activity and chemical stability of incorporated active substances.

When organoleptic changes occur and makes muscle foods unacceptable to the consumer, meat loss quality is considered spoiled. Contamination by microorganisms is of the main causes for organoleptic spoilage. Food package contributes to an easier distribution and protects food from environmental conditions, such as light, oxygen, moisture, microorganisms, mechanical injuries, and dust. Through the application of active packaging systems, these conditions can be regulated in several ways and, depending on the requirements of the packaged food, food deterioration can be significantly reduced.

Antimicrobial packaging (AP) is a type of active packaging and represents a promising solution, especially tailored to improve safety and to delay spoilage. Antimicrobials can be coated, incorporated, immobilized, or surface modified onto package materials. The AP development is limited, due to availability of antimicrobials and new polymer materials, regulatory concerns, and appropriate testing methods. Future work must focus on the use of biologically active derived antimicrobial compounds bound to polymers. The need for new antimicrobials with wide spectrum of activity and low toxicity will increase [22].

Plant-derived extracts (PDE) represent good candidates for antimicrobial packaging. It is possible that research and development of 'intelligent' or 'smart' antimicrobial packages will follow. These will be materials that sense the presence of microorganism in the food, triggering antimicrobial mechanisms, as a response in a controlled manner. Success of AP technologies for food applications is related to participation and collaboration of research institutions, industry, and government regulatory agencies.

The need to use materials more sustainable and more compatible with food represents a new market and leads to an intense activity in the study of natural substances for the production of biodegradable wrapping and edible coatings. The exploration of plant-derived antimicrobials should be an innovative way to find new alternative substances for food preservation via active packaging. Furthermore, the use of natural antimicrobial products derived from plant is important because they are tolerable for the consumer. The exploration of plant-derived antimicrobials represents an innovative way to find new alternative substances as food preservatives for active packaging [23].

The idea is that the exploration of PDE as preservatives can provide an innovative way to find new alternative substances for meat preservation. The use of PDE as antimicrobial has been already reported and is important since they represent a lower perceived risk to the consumer as well consumer's demand for minimally processed, preservative free products increases. To be suitable, the antimicrobial PDE should be cheap, ecologically acceptable, and target tailored, besides being effective [24, 25].

The International Life Sciences Institute-Europe produced a comprehensive document on the use of plant materials in food products [26]. The report stresses that the ingredient for use in food products must be well identified and characterized. Accurately identification of the starting material needs in order to ensure that the plant materials for food use are consistent with respect to quality and quantity of active ingredient. The method of preparation must meet good manufacturing practices. Antimicrobial preservative releasers (films) of organic acids, as sorbic acid, silver zeolite, spice and herb extracts, allylisothiocyanate, enzymes, for example, lysozyme and bacteriocins are used for growth inhibition of spoilage and pathogenic bacteria.

6. New packaging technologies and the development of food packaging materials and nanotechnologies with food/feed applications

With the aim of development of active and passive materials for the use in the design of packages, coatings, and packaging technologies that help maintain and improve the sensorial and nutritional characteristics and safety of foodstuffs, as well as to increase their shelf life, nanofabrication technologies are emerging as valuable solutions. In this way, nanostructures and encapsulations for food applications, based on renewable materials, either edible or inedible, can be obtained, which possess active and bioactive properties to improve and develop preservation and packaging processes of foods and/or their ingredients.

Edible films are defined as a layer of material, which can be also eaten, yet provides a barrier to moisture, oxygen, and solute movement for the food [27]. If the films are not eaten, they could become biodegradable in the environment.

In recent years, the interest in microbial extracellular polysaccharides has increased, as they are candidates for many commercial applications in different industrial sector like food and pharmaceuticals. Esopolysaccharides are natural, non-toxic, and biodegradable polymers that, besides the interest on their application in the health and biotechnology, are used as stabilizers, gelling agents, and thickeners in food and cosmetic industries. Synthesis of value-added biochemical from biomass using microorganism are a promising alternative. Microbial extracellular polysaccharides cannot find its proper place in the market, unless it can be produced economically.

The microorganisms producing EPS are mainly bacteria belonging to the species *Xanthomonas, Leuconostoc, Sphingomonas, Alcaligenes,* and many other, which produce xanthan dextran, gelling, and curdling, all known to have industrial applications. However, when compared with the synthetic polymers, microbial EPS represent a small sector in the market because of their costly production processes [28–30]. This could be avoided by developing cost-effective and environmentally friendly production processes, such as investigating the potential use of cheaper fermentation substrates as agro-industrial and agricultural by-products and waste contributing also to reduce their environmental concern as, for instance, Olive Mill Wastewater and Pomace in the Mediterranean area [31]. Proper pre-treatment for both substrata have been developed to eliminate undesirable constituents with antimicrobial activity as phenols in oil mill water [32–34].

The fungus *Aureobasidium pullulans* produces pullulan, an extracellular and neutral polysaccharide, that is, a linear polymer mainly consisting of malt triose units interconnected to each other by α-(1,6)-glycoside bonds. This confers a good solubility in water [35]. Pullulan is listed as Existing Food Additives. The Japan Food Chemical Research Foundation and the Joint Food and Agriculture Organization/World Health Organization Expert Committee on Food Additives (JECFA) approved it as Generally Recognized as Safe (GRAS, E1204) in 2002 [36].

The incorporation of natural antimicrobial substances into edible films is of big concern in recent years because it can enhance the safety and quality of food products by controlling or reducing the growth of food-borne and spoilage microorganisms.

7. Molecular biology key role

Among the molecular biology techniques, PCR and multiplex PCR of DNA are very useful to specifically identify and detect cells of microorganisms affecting food quality, but they are not able to distinguish live from dead cells. The use of PMA™ dye is a valid alternative method to assess bacterial growth or its inhibition instead of the colorimetric methods using a tetrazolium salt [37, 38]. Furthermore, tetrazolium salts are not suitable to assess the growth of microaerophilic bacteria [39]. Other techniques as real-time polymerase chain reaction (qPCR) could also be used, but their uses imply equipment which cost is 10-fold higher than that of a thermal cycler and higher specialized knowledge to perfect protocols for its use. Instead, PCR and multiplex PCR are cheaper, less time consuming, and the results are easily readable in comparison with qPCR and microbiological methods. All that let to perform rapid and efficacious massive screening of samples as well as safety control survey. In addition, molecular biology techniques let correct species and strain identification within mixed microbial populations associated to foods and develops databases for the food source assignment of microbes.

They let to detect food-borne microbes and assess the risk they represent, comprising of viable or infectious agents in non-cultivable states, too.

However, thanks to the metagenomics technology, nowadays research focus on the whole food-associated microbial populations. They are studied at different taxonomic levels. Food microbiota is revealed with rRNA amplicon-based high-throughput sequencing approaches for food quality screening, monitoring population dynamics and meta-analyses based on food microbiota interactive databases. This allows investigating the different groups of food-associated microorganisms, mainly pathogens, spoilers, and fermentation player, as well as their interactions and influence on food quality.

8. The neem opportunity

In recent years, a major concern emerges in the field of bacterial infections to stimulate the development of innovative molecules with antimicrobial activity. Nowadays, zoonotic food- and water-borne pathogens are becoming more resistant to antibiotics. Many resistant strains have been isolated from food and could be entering the human gastrointestinal tract on an almost daily basis. The increasing incidence of food-borne diseases, coupled with the resultant social and economic implications, causes a constant striving to produce safer feed and food as to develop new natural antimicrobial agents. Plants and their agro-industrial waste and by-products represent sources of biologically active substances as potential antibiotics. According to World Health Organization [24], medicinal plants would be the best source to obtain a variety of useful drugs. Many plants produce secondary metabolites, which act against wound-contaminating bacteria and parasites [40].

Nowadays, among the most promising emerging species, Neem (*Azadirachta indica* A. Juss) is considered an effective source of environmentally powerful natural products. It is believed to be one of the most promising trees of the twenty-first century for its great potential in pest management, environmental protection, and medicine. US EPA tested biocidal efficacy and absence of environmental negative impact of neem products [41]. Among the many products obtained from the tree, neem oil (NO) is the most commercially relevant (www.organic-neem.com/why_parker_neem.html). NO is obtained by mechanical extraction of the kernels. Actually, NO is mainly utilized as natural insecticide, whereas the resulting residue, known as neem cake (NC), is used in agriculture as fertilizer or as animal feed. NCE (Neem Cake Extract) was a selected model, due to its low cost and the antimicrobial potentiality of neem, for exploring antibacterial with a view to mass treatment of meat products. NO shows strong antimicrobial activity against different microbial populations from food (spoilage microorganism and food-borne pathogens) and wound-contaminating bacteria [42–51]. The low cost and the available quantity of NC make it a potentially important raw material for developing new eco-friendly insecticidal products [52–61].

An important aspect concerns the variability of neem products, whose chemical composition must be determined. The metabolome determination of neem products was obtained using high performance thin-layer chromatography (HPTLC). HPTLC, the last evolution of planar chromatography, allows one to detect the majority of the constituents of an extract in an identifying track, named the fingerprint. An application of the HPTLC fingerprint method was developed as specific application to determine the identification of composition of utilized plants as complex extracts and their qualitative and quantitative pattern, in order to maintain the correspondence between composition and activity. It is necessary to standardize natural products, in order to establish the scientific evidences of their security and biological activity. In fact, the metabolomics approach allows obtaining the widest possible coverage, in terms of the type and number of analyzed compounds. The fingerprint by HPTLC method was successfully used to determine the herbal composition of neem studied product [62–69], evidencing the complexity in composition and the multiplicity of activities.

The antibacterial activity of neem products is known to be mainly focused on the antimicrobial properties of azadirachtin A. However, azadirachtins are the main constituents in neem oil and salannin is predominant in neem cake. The study on the metabolomic and biological activity of neem products allow us to understand the importance to handle the whole phytocomplex. Based on NO antibacterial activity, it should have several field relating human and animal health and wellbeing as well as feed and food preservation.

9. Conclusions

New packaging technologies and the development of food packaging materials and nanotechnologies with food/feed applications are an important front to produce food in a way that ensure it is used more efficiently and equitably, to reduce emerging/developed-country food waste, energy cost along FSC, the environment impact of agricultural, and agro-industrial products along FSC. Main objectives to be pursued are the development of innovative solutions in food packaging for sustainable production processes and innovative packaging systems 'green label' by the use of biodegradable and recyclable films with properties for containing antimicrobials to control the microbial contamination and food spoilage by PDE as new preservatives and food contaminant.

New promising methods can be also considered for the utilization of organic biodegradable innovative materials, like chitosan, fern, algae, and others, in order to avoid the utilization of metals, that is, silver or gold, and limit the use of pollutant plastics in the inclusion in the texture [70–78]. After the results obtained on the antimicrobial activity of NO and NCE, further researches are in progress to develop smart packaging by nanotechnology microencapsulation, as already obtained for insecticidal activity [79–81].

Author details

Marcello Nicoletti[1] and Paola Del Serrone[2*]

*Address all correspondence to: paola.delserrone@cra.it

1 Department of Environmental Biology, Sapienza University of Rome, Rome, Italy

2 Council of Agricultural Research and Economics (CREA), Zootechnics and Aquaculture Research Center, Monterotondo, Italy

References

[1] European Commission. Agriculture, Forestry and Fisheries Statistics. Luxembourg: Publications Office of the European Union; 2013. 249 pp. DOI: 10.2785/45595

[2] FAO. Mapping supply and demand for animal-source foods to 2030. In: Robinson TP, Pozzi F, editors. Animal Production and Health Working Paper No. 2. Rome; 2011. 141 pp

[3] Brockgreitens J, Abbas A. Responsive food packaging: Recent progress and technological prospects. Comprehensive Reviews in Food Science and Food Safety. 2016;**5**:3-115. DOI: 10.1111/1541-4337.12174

[4] Lee SY, Lee SJ, Choi DS, Hur SJ. Current topics in active and intelligent food packaging for preservation of fresh foods. Journal of the Science of Food and Agriculture. 2015;**95**(14):2799-810. DOI: 10.1002/jsfa.7218

[5] Vera P, Echegoyen Y, Canellas E, Nerín C, Palomo M, Madrid Y, Cámara C. Nano selenium as antioxidant agent in a multilayer food packaging material. Analytical and Bioanalytical Chemistry. 2016;**408**(24):6659

[6] Otoni CG, Espitia PJP, Avena-Bustillos RJ, McHugh TH. Trends in antimicrobial food packaging systems: Emitting sachets and absorbent pads. Food Research International. 2016;**83**:60

[7] Kim JU, Ghafoor K, Ahn J, Shin S, Lee SH, Shahbaz H M, Shin H-H, Kim S, Park J. Kinetic modelling and characterization of a diffusion-based time-temperature indicator (TTI) for monitoring microbial quality of non-pasteurized angelica juice. LWT—Food Science and Technology. 2016;**67**:143

[8] Lundqvist J, de Fraiture C, Molden D. Saving water: From field to fork-curbing losses and wastage in the food chain. In: SIWI Policy Brief. Stockholm, Sweden: SIWI; 2008

[9] Kader AA. Increasing food availability by reducing postharvest losses of fresh produce. Acta Horticulturae. 2005;**682**:2169-2175

[10] Choudhury ML. Recent developments in reducing post-harvest losses in the Asia-Pacific region. In: Rolle RS, editor. Reports of the APO Seminar on Reduction of Postharvest Losses of Fruit and Vegetables; October 2004. Tokyo, Japan: FAO, Asian Productivity Organization (APO); 2006. pp. 5-11

[11] HLPE. Food losses and waste in the context of sustainable food systems. A report by the High Level Panel of Experts on Food Security and Nutrition of the Committee on World Food Security. Rome; 2014. www.fao.org/cfs/cfs-hlpe

[12] Nellemann C et al, editors. The Environmental Food Crisis. Kenya: United Nations Environment Programme, Nairobi; 2009

[13] Meat Industry Service. Natural Antimicrobials, Parasitic Bacteria and Bacteriophages; 2006. http://www.meatupdate.csiro.au/new/Natural%20Antimicrobials,%20Parasitic%2Bacteria%20and%20Bacteriophages.pdf [Accessed: 19 December 2014]

[14] Maruchecka A, Greis N, Menac C, Cai L. Product safety and security in the global supply chain: Issues, challenges and research opportunities. Journal of Operations Management. 2011;**29**:707-720. DOI: 10.1016/j.jom.2011.06.007

[15] Thakur MS, Ragavan KV. Biosensors in food processing. Journal of Food Science and Technology. 2013;**50**(4):625-641. DOI: 10.1007/s13197-012-0783-z

[16] Singh RK, Singh N. Quality of packaged foods. In: Han JH, editor, Innovations in Food Packaging. Amsterdam: Elsevier Academic Press; 2005. pp. 22-24

[17] De Kruijf N, Van Beest M, Rijk R, Sipiläinen-Malm T, Paseiro Losada P, De Meulenaer B. Active and intelligent packaging: Applications and regulatory aspects. Food Additives & Contaminants. 2002;**19**:144-162

[18] Kerry JP, O'Grady MN, Hogan SA. Past, current and potential utilization of active and intelligent packaging systems for meat and muscle-based products: A review. Meat Science. 2006;**74**:113-130

[19] Cooksey K. Effectiveness of antimicrobial food packaging materials. Food Additives & Contaminants. 2005;**22**:980-987

[20] Ozdemir M, Floros JD. Active food packaging technologies. Critical Reviews: Food Science and Nutrition. 2004;**44**(3):185-193

[21] Coma V. Bioactive packaging technologies for extended shelf life of meat-based products. Meat Sci. 2008;**78**:90-103.

[22] Malhotra B, Keshwani A, Kharkwa H. Antimicrobial food packaging: Potential and pitfalls. Frontiers in Microbiology. 2015;**6**:611. DOI: 10.3389/fmicb.2015.00611. PMCID: PMC 4468856

[23] Dobrucka R, Cierpiszewski R. Active and intelligent packaging food—Research and development—A review. Polish Journal of Food and Nutrition Sciences. 2014;**64**(1):7-15. DOI: 10.2478/v10222-012-0091-3

[24] WHO, World Health Organization. Quality Control Methods for Medicinal Plant Materials. Geneva: WHO Library Cataloguing; 2011

[25] Negi PS. Plant extracts for the control of bacterial growth: Efficacy, stability and safety issues for food application. International Journal of Food Microbiology. 2012;**156**(1):7-17. DOI: 10.1016/j.ijfoodmicro.2012.03.006

[26] Schiltera B, Anderssonb C, Antonc R, Constablea A, Kleinerd J, O'Briene J, Renwickf AG, Korverg O, Smith F, Walker I. Guidance for the safety assessment of botanicals and botanical preparations for use in food and food supplements. Food and Chemical Toxicology. 2003;**41**:1625-1649. DOI: 10.1016/S0278-6915(03)00221-7

[27] Bourtoom T. Edible films and coatings: Characteristics and properties (review article). International Food Research Journal. 2008;**15**:237-248

[28] Aguilera M, Quesada MT, Aguila VG, Morillo JA, Rivadeneyra MA, Romos-Cormenzana A, Monteoliva-Sanchez M. Characterization of *Paenibacillus jamilae* strains that produce exopolysaccharide during growth on and detoxification of olive mill wastewaters. Bioresource Technology. 2008;**99**:5640-5644

[29] Ahvenainen R. Active and intelligent packaging: An introduction. In: Ahvenainen R, editor. Novel Food Packaging Techniques. Cambridge, UK: Woodhead Publishing Ltd; 2003. pp. 5-21

[30] Appendini P, Hotchkiss JH. Review of antimicrobial food packaging. Innovative Food Science and Emerging Technologies. 2002;**3**:113-121

[31] Stasinakis SA, Elia I, Petalas AV, Halvadakis CP. Removal of total phenols from olive-mill wastewater using an agricultural by-product, Olive Pomace. Journal of Hazardous Materials. 2008;**160**(2-3):408-413. DOI: 10.1016/j.jhazmat.2008.03.012

[32] Sutherland IW. Bacterial exopolysaccharides. In: Kamerling JP, editor. Comprehensive Glycoscience. Amsterdam: Elsevier; 2007

[33] Nicolaus B, Kambourova M, Toksoy Öner E. Exopolysaccharides from extremophiles: From fundamentals to biotechnology. Environmental Technology. 2010;**31**:1145-1158

[34] Salehizadeh H, van Loosdrecht MCM. Production of polyhydroxyalkanoates by mixed culture: Recent trends and biotechnological importance. Biotechnology Advances. 2004;**22**:261-279

[35] Xiao Q, Lim, LT, Tong Q. Properties of pullulan-based blend films as affected by alginate content and relative humidity. Carbohydrate Polymers. 2012;**87**:227-234

[36] Chaen H. Pullulan. In: Imeson A, editor. Food Stabilisers. Thickeners and Gelling Agents. West Sussex, UK: Blackwell Publishing; 2010. pp. 266-274

[37] Grare M, Fontanay S, Cornil C, Finance C, Duval RE. Tetrazolium salts for MIC determination in microplates: Why? Which salt to select? How? Journal of Microbiological Methods . 2008;**75**:156-159

[38] Rahman M, Kuhn M, Rahman P, Olsson-Liljequist B, Molby R. Evaluation of a scanner-assisted. Colorimetric MIC method for susceptibility testing of Gram-negative fermentative bacteria. Applied and Environmental Microbiology. 2004;**70**:2398-2403

[39] Klančnik A, Piskernik S, Jeršek B, Smole MS. Evaluation of diffusion and dilution methods to determine the antibacterial activity of plant extracts. Journal of Microbiological Methods. 2010;**81**:121-126

[40] Swati D, Kumar SP, Jyoti R, Renu T, Arti B. Phytochemical analysis of seeds of certain medicinal plants. International Research Journal of Pharmacy. 2014;**5**(2):102-105. DOI: 10.7897/2230-8407.050221

[41] EPA. Cold Pressed Neem Oil PC Code 02500. U.S. Environmental Protection Agency Office of Pesticide Programs. Biopesticides and Pollution Prevention Division; Washington, D.C. 2012

[42] Del Serrone P, Nicoletti M. Antimicrobial activity of a neem cake extract in a broth model meat system. International Journal of Environmental Research and Public Health. 2013;**10**:3282-3295

[43] Del Serrone P, Nicoletti M. Alternative control of muscle spoilage by a neem cake extract. Pharmacologyonline Archive. 2013;**3**:132-136

[44] Del Serrone P, Nicoletti M. Impiego di un estratto di neem cake per la conservazione della carne fresca confezionata. Igiene, Alimenti, Disinfestazione e Igiene Ambientale. 2013;**30**(4):4-8

[45] Del Serrone P, Failla S, Nicoletti M. Natural control of bacteria affecting meat quality by a neem (*Azadiracta indica* A Juss) cake extract. Natural Product Research. 2014;**29**(10):985-987. DOI: 10.1080/14786419.2014.964708

[46] Del Serrone P, Nicoletti M 2014. Evaluation of a mono-component and a multi-component herbal extracts as candidates for antimicrobial packaging of fresh retail meat. In: Cimmino S, Pezzuto M, Silvestre C, editors. Proceedings: Eco-sustainable Food

Packaging based on Polymer nanomaterials. International Conference COST ACTION FA0904; 26-28 February 2014; CNR Rome, Italy. p. 36

[47] Del Serrone P, Toniolo C, Nicoletti M. Neem (*Azadirachta indica* A. Juss) seed oil a natural preservative to control meat spoilage. Foods. 2015;4:3-14. DOI: 10.3390/foods4010003

[48] Del Serrone P, Toniolo C, Nicoletti M. Neem (*Azadirachta indica* A. Juss) seed oil to tackle enteropathogenic *Escherichia coli*. BioMed Research International. 2015;**343610**:10 pp. DOI: 10.1155/2015/343610

[49] Del Serrone P, Toniolo C, Nicoletti M. Evaluation of a three-component plant derived extract as preservative for fresh retail meat. Research & Reviews: Journal of Food Dairy Technology. 2015;**3**(3):24-31

[50] Del Serrone P, Toniolo C, Nicoletti M. Assessment of a garlic product's antibacterial activity and its effects towards enteropathogenic *Escherichia coli*. International Journal of Food Sciences and Nutrition. 2016;**1**(3):58-66

[51] Del Serrone P, Buttazzoni L, Nicoletti M. Nutrition and multiresistance alert. EC Nutrition. 2016;**4**(1):772-783

[52] Del Serrone P, De Matteis G, Rufrano D, Veneziano V, Claps S, et al. Goat wound-contaminating-bacteria restraint by neem oil. SOJ Microbiology & Infectious Diseases. 2015;**3**(2):1-10

[53] Benelli G, Nicoletti M. Shedding light on bioactivity of botanical by-products: Neem cake compounds deter oviposition of the arbovirus vector *Aedes albopictus* (Diptera: Culicidae) in the field. Parasitology Research. 2013;**110**:2013-2021

[54] Benneli G, Canale A, Toniolo C, Higuchi A, Murugan K; Pavela R, Nicoletti M. Neem (*Azadirachta indica*): Towards the ideal insecticide? Natural Product Research. 2016;1-18

[55] Benelli G, Bedini S, Cosci F, Toniolo C, Conti B, Nicoletti M. Larvicidal and ovideterrent properties of neem oil and fractions against the filariasis vector *Aedes albopictus* (Diptera: Culicidae): A bioactivity survey across production sites. Parasitology Research. 2015;**114**(1):227-236

[56] Benelli G, Murugan K, Panneerselvam C, Madhiyazhagan P, Conti B, Nicoletti M. Old ingredients for a new recipe? Neem cake, a low-cost botanical by-product in the fight against mosquito-borne diseases. Parasitology Research 2015;**114**(2):391-397. DOI: 10.1007/s00436-014-4286-x

[57] Chandramohan B, Murugan K, Madhiyazhagan P, Kovendan K, Mahesh Kumar P, Panneerselvam C, Dinesh D, Subramaniam J, Rajaganesh R, Nicoletti M, Canale A, Benelli G. Neem by-products in the fight against mosquito-borne diseases: Biotoxicity of neem cake fractions towards the rural malaria vector *Anopheles culicifacies* (Diptera: Culicidae). Asian Pacific Journal of Tropical Biomedicine. 2016;**6**(6):472-476. DOI: 10.1016/j.apjtb.2015.11.013

[58] Nicoletti M, Maccioni O, Ciocchetti T, Mariani S, Vitali F. Neem tree (*Azadirachta indica* A. Juss) as source of bioinsectides. In: Perveen F, editor. Insecticides—Advances in Integrated Pest Management; 2012. ISBN: 978-953-307-780-2. http://www.intechopen.com/articles/show/title/neem-tree-azadirachta-indica-a-juss-as-source-of-bioinsectides

[59] Nicoletti M, Serafini M, Aliboni A, Armando D'Andrea A, Mariani S. Toxic effects of neem cake extracts on *Aedes albopictus* (Skuse) larvae. Parasitology Research. 2010;**107**:89-94

[60] Nicoletti M, Murugan K, Benelli G. Chapter 11 Emerging insect-borne diseases not-involving the human health. In: Trdan S, editor. Insecticide Resistance. Rijwcka, Croatia: In-Tech; 2016. pp. 219-242

[61] Nicoletti M, Murugan K, Canale A, Benelli G. Neem-borne molecules as eco-friendly tools against vectors of economic importance. Current Organic Chemistry. 2016;**20**(25):2681-2689. DOI: 10.2174/1385272820666160218233923

[62] Nicoletti M. HPTLC fingerprint: a modern approach for the analytical determination of botanicals. Revista Brasileira de Farmacognosia. 2011;**21**(5):818-823

[63] Nicoletti M, Toniolo C. HPTLC fingerprint analysis of plant staminal cell products. Journal of Chromatography & Separation Techniques. 2012;**3**:7-11. DOI: 10.4172/2157-7064.1000148

[64] Nicoletti M, Toniolo C, Gallo FR, Multari G, Palazzino G. Traceability in multi-ingredient botanicals by HPTLC fingerprint approach. Journal of Planar Chromatography. 2013;**26**(3):243-247

[65] Nicoletti M. Phytochemical techniques in complex botanicals. The XXI Century Analytical Technique. Journal of Chromatography & Separation Techniques. 2012;**4**:186. DOI: 10.4172/2157-7064.1000186

[66] Gallo FR, Multari G, Federici E, Palazzino G, Nicoletti M, Petitto V. The modern analytical determination of Botanicals and similar novel natural products by the HPTLC Fingerprint approach. In: Rahman AU, editor. Studies in Natural Products Chemistry. Elsevier; 2012. **37**; pp. 217-258

[67] Nicoletti M, Toniolo C. Fingerprint HPTLC analysis of marketed products. Pharmacology. 2013;**3**(6):30-36

[68] Toniolo C, Nicoletti M, Maggi F, Venditti A. Determination by HPTLC of chemical composition variability in raw material used in botanicals. Natural Products Research. 2014;**28**(2):119-126

[69] Nicoletti M, Toniolo C. Analysis of multi-ingredient food supplements by fingerprint HPTLC approach. Journal of Chemistry and Chemical Engineering. 2015;**9**:239-244

[70] Anbu P, Murugan K, Madhiyazhagan P, Dinesh D, Subramaniam J, Panneerselvam C, Suresh U, Alarfaj A A, Murugan K, Munusamy A., Higuchi A, Hwang J-S, Kumar S, Nicoletti M, Benelli G. Green-synthesised nanoparticles from Melia azedarach seeds and the cyclopoid

crustacean *Cyclops vernalis*: an eco-friendly route to control the malaria vector *Anopheles stephensi*?, Nat Product Res 2016. 30:18, 2077-2084, DOI: 10.1080/14786419.2015.1114935 http://dx.doi.org/10.1080/14786419.2015.1114935

[71] Panneerselvam C, Murugan K, Roni M, Tabini A, Suresh U, Madhiyazhagan P, Subramanian J, Chandramohan B, Dinesh D, Kovendan K, Mahesh KP, Nicoletti M, Higuchi A, Kumar S, Benelli G. Fern-synthetized nanoparticles in the fight against malaria: LC/MS analysis of *Pteridium aquilium* leaf extract and biosynthesis of silver nanoparticles with high mosquicidal and antiplasmodial activity. Parasitology Research. 2016;**115**(3):997-1013. DOI: 10.1007/s00436-015-4828-x

[72] Sujitha V, Murugan K, Paulpandi M, Panneerselvam C, Suresh U, Roni M, Nicoletti M, Higuchi A, Madhiyazhagan P, Subramaniam J, Dinesh D, Vadivalagan C, Chandramohan B, Alarfaj AA, Munusamy MA, Barnard DR, Benelli G. Green-synthesized silver nanoparticles as a novel control tool against dengue virus (DEN-2) and its primary vector Aedes aegypti. Parasitology Research. 2015.**114**(9):3315-25. DOI: 10.1007/s00436-015-4556-2

[73] Murugan K, Samidoss CM, Panneerselvam C, Higuchi A, Roni M, Suresh U, Chandramohan B, Subramaniam J, Madhiyazhagan P, Dinesh D, Rajaganesh R, Alarfaj AA, Nicoletti M, Kumar S, Wei H, Canale A, Mehlhorn H, Benelli G. Seaweed-synthesized silver nanoparticles: An eco-friendly tool in the fight against *Plasmodium falciparum* and its vector *Anopheles stephensi*? Parasitology Research. 2015;**114**(11):4087-4097. DOI: 10.1007/s00436-015-4638-1

[74] Jaganathan A, Murugan K, Panneerselvam C, Madhiyazhagan P, Dinesh D, Vadivalagan C, Aziz AT, Chandramohan B, Suresh U, Rajaganesh R, Subramaniam J, Nicoletti M, Higuchi A, Alarfaj AA, Munusamy MA, Kumar S, Benelli G. Earthworm-mediated synthesis of silver nanoparticles: A potent tool against hepatocellular carcinoma, *Plasmodium falciparum* parasites and malaria mosquitoes. Parasitology Research. 2016;**65**(3):276-284. DOI: 10.1016/j.parint.2016.02.003

[75] Muruga K, Panneerselvam C, Subramaniam J, Madhiyazhagan P, Hwang J-S, Wang L, Dinesh D, Suresh U, Roni M, Higuchi A, Nicoletti M, Benelli G. Eco-friendly drugs from the marine environment: spongeweed-synthesized silver nanoparticles are highly effective on Plasmodium falciparum and its vector Anopheles stephensi, with little non-target effects on predatory copepods. Environ Sci Pollut Res 2016. **23**:16671-16685. DOI 10.1007/s11356-016-6832-9

[76] Murugan K, Aamina JM, Paarnelservam C, Dinesh D, Suresh U, Subrapamanian Madhiyazaghan P, Hwang JS, Wang L, Nicoletti M, Benelli G. *Aristolochia indica* green-synthesized silver nanoparticles: a sustainable control tool against the malaria vector *Anopheles stephensi*? Research in Veterinary Science. 2015;**102**:127-135

[77] Amerasan D, Natarej T, Murugan K, Madhiyazhadan P, Panneerselvan C, Nicoletti M, Benelli G. Myco-synthesis of silver nanoparticles using *Metarhizium anisopliae* against the rural malaria vector *Anopheles culicifacies* Giles. Journal of Pest Science. 2015;**9**(1):249-256. DOI: 10.1007/s10340-0.15-0675-x

[78] Benelli G. Plant-mediated synthesis of nanoparticles: A newer and safer tool against mosquito-borne diseases? Asian Pacific Journal of Tropical Biomedicine. 2016;**6**(4):353-354

[79] Chandramohan B, Murugan K, Panneerselvam C, Madhiyazhagan P, Chandirasekar R, Dinesh D, Kumar PM, Kovendan K, Suresh U, Subramaniam J, Rajaganesh R, Aziz AT, Syuhei B, Alsalhi MS, Devanesan S, Nicoletti M, Wei H, Benelli G. Characterization and mosquitocidal potential of neem cake-synthesized silver nanoparticles: Genotoxicity and impact on predation efficiency of mosquito natural enemies. Parasitology Research. 2016;**115**(3):1015-25. DOI: 10.1007/s00436-015-4829-9

[80] Vijayan V, Aafreen S, Sakthivel S, Reddy RK. Formulation and characterisation of solid lipid nanoparticles loaded neem oil for topical treatment of acne. Journal of Acute Disease. 2013;**2**(4):282-286. DOI:10.1016/S2221-6189(13)60144-4

[81] Rajendran R, Radhai R, Balakumar C, Ahamed HAM, Vigneswaran C, Vaideki K. Synthesis and characterization of neem chitosan nanocomposites for development of antimicrobial cotton textiles. Journal of Engineered Fibers and Fabrics. 2012;**7**(1):136-141

Chapter 9

Novel Soft Meals Developed by 3D Printing

Mai Kodama, Yumiko Takita, Hideaki Tamate,
Azusa Saito, Jin Gong, Masato Makino, Ajit Khosla,
Masaru Kawakami and Hidemitsu Furukawa

Additional information is available at the end of the chapter

http://dx.doi.org/10.5772/intechopen.70652

Abstract

Recently, 3D printing is being applied to various fields. 3D printing of foods has been developed; however, there are many challenges. To overcome the challenges, we have started a new research group named "Yonezawa Itadakimasu Research Group," to focus on the development of 3D printing applications for manufacturing food. We have developed Novel jelly foods that are shaped by 3D printed molds. Fused deposition modeling (FDM) 3D printer for food manufacturing makes the 3D printed molds. First step of making 3D printing mold is to print a cast. Then, food grade silicone is poured into the cast to make a mold. This type of 3D printed mold can be used widely, such as making sweets, restaurant menus, and care foods by changing the design depending on the use of application. Secondly, we started to develop 3D food printers. This type of challenge to develop future foods by 3D printing technology may have a major impact on the care food because the looks of foods are important and will be improved by 3D printing.

Keywords: 3D printer, food printer, FDM, care food, mold

1. Introduction

3D printing finds its application in medicine, manufacturing, sensors, micro-and nanotechnology, and custom-made art and design. Most widely used application of 3D printing is in industrial equipment such as production of prototype parts of industrial products and confirmation of shape. Over the last decade, application of 3D printers has increased, and free-formable 3D printers are becoming widely used. Now 3D printers can be purchased from electronic retail stores to online retail stores such as Amazon. Many low-cost compact and portable 3D printers have been developed, but large-scale 3D printers that can output furniture to homes have also been developed. Varieties of 3D printers exist, such as 3D printing of foods, medicines, and organs. Two distinct types of Product Data Management (PDM) printers are being developed for 3D printing of foods: (1) Fused deposition modeling (FDM) type of 3D food printers that directly output foods; and (2) FDM type which makes 3D printing mold, i.e., to print a cast for food molding.

To increase the awareness and use of 3D printers, we have set up locations where ordinary people can use 3D printers such as "Yonezawa Itadakimasu Research Group" and "Eki Fab." Yonezawa Itadakimasu Research Group is researching and developing new foods utilizing 3D printers. Eki Fab is a fabrication space inside the station building, and anyone can freely use a 3D printer. We have held food events at the Eki Fab.

This paper describes the principle of 3D printers, application of 3D printing in food manufacturing, "Yonezawa Itadakimasu Research Group," "Eki Fab," and the future of foods made by 3D printer.

2. Principle of 3D printer

3D printer is a device for manufacturing a three-dimensional object that stacks materials based on 3D data. 3D printers can be roughly classified into lamination methods such as fused deposition modeling (FDM), powder lamination method, inkjet and bathtub stereolithography (**Figure 1**).

- Fused deposition modeling (FDM)

FDM is the most inexpensive and major modeling method as a 3D printer for public because patent has expired. In the FDM 3D printer, a string-like resin called a filament is melted by a heater and discharged from a nozzle, so that it is shaped. It is necessary to introduce support structure because structures floating in the air cannot be shaped in principle of modeling. The main resins currently used are various resins refer to thermoplastic ABS and PLA. ABS is a resin obtained by copolymerizing acrylonitrile, butadiene, and styrene in general. In addition, the use of various resins such as super engineering plastics such as is beginning to be studied such as polycarbonate (PC) and polyphenol sulfone (PPSU).

As an example of application to foods, FDM type printers are used to make food molds such as chocolate and jelly molds.

- Powder lamination method

In the powder lamination method, the surface of a bed containing powder is thinly compacted at an arbitrary place, and this is repeatedly laminated as one layer. It can handle a wide range of materials such as metal powder, resin powder, and ceramic powder. It is an expensive device which is difficult to purchase for individuals because it has higher modeling precision than FDM. Also, since the material is powdery, management and safety considerations are necessary. This method is further classified into two methods: selective laser sintering, which locally heats with laser to melt and consolidate powder, and powder fixing method that fixes with adhesive.

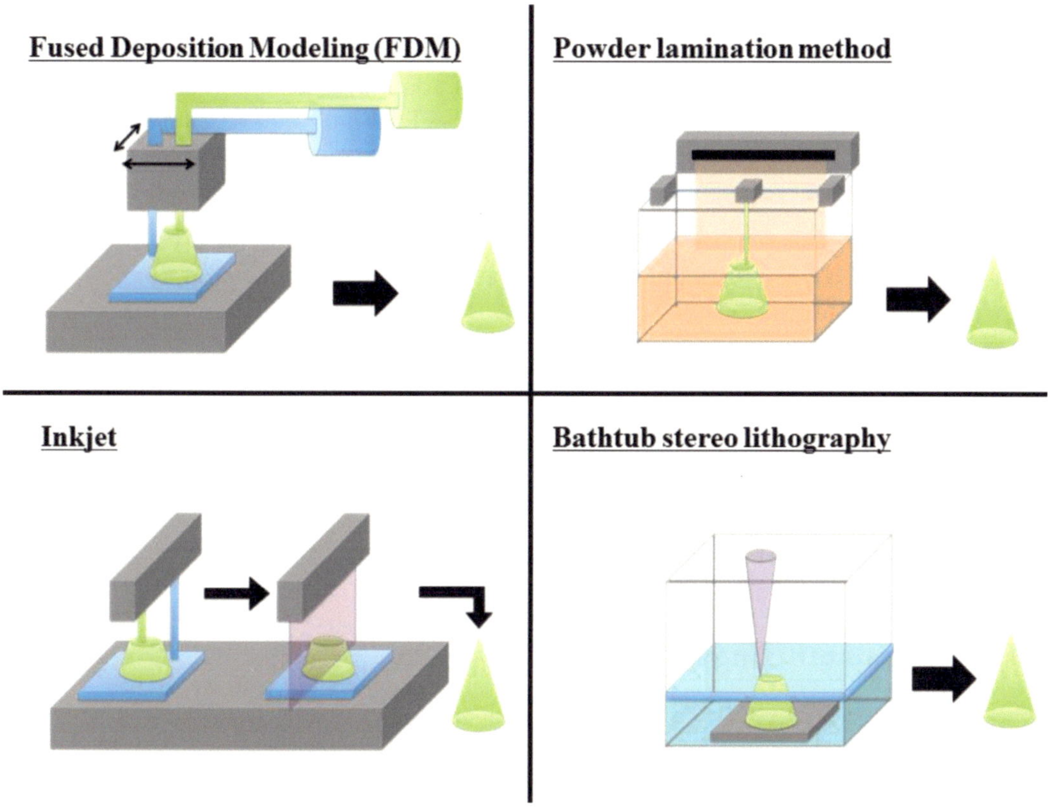

Figure 1. Modeling methods of 3D printer.

Powder sintering method is like FDM, but among thermoplastic resins, powder of crystalline polymer such as nylon 11, 12, polyamide 11, 12 is mainly used, and metal powders and ceramics powder can also be handled. Powder is a support structure; hence, there is no need to design the support structure.

Powder fixing method is chemical energy is given to ceramics powder such as gypsum, and this chemically combines the powders and solidifies. As an example, calcium sulfate causes a hydration reaction when water is added, it solidifies into dihydrate gypsum. Full color printing is also possible if coloring is charged in water. In addition, there are cases in which polymer is sprayed on natural sand or artificial sand as an adhesive. Particularly, 3D printers for producing sand mold for casting using furan type or phenol type resins become conspicuous in the market. This powder fixing method is sometimes called inkjet, described later, because liquid (ink) is ejected from a nozzle.

- Inkjet

As a 2D printer, a method of ejecting ink from a nozzle and solidifying the material is called an inkjet lamination method. The point that the fine particles of the photo-cross-linkable resin used as a material is solidified by irradiating ultraviolet rays is different from the powder fixing method. Thickness of a layer stacked by a 3D printer is called stacking pitch. While the stacking pitch of a typical FDM is 0.2 mm, the stacking pitch of inkjet is as fine as 0.01 mm, so it becomes a beautiful and smooth shaped object. Materials often used for inkjet are urethane acrylate type (acrylic resin) of radical polymerization. Photo-cross-linkable resins are important for ink control to determine shaping accuracy. Since the reaction rate is also important, radical polymerization with a high polymerization rate is used.

As an example of application to foods, there are cases where dispensers are used to serve food in a restaurant or to make pizza and utilization in events.

- Bathtub stereolithography

Bathtub stereolithography is a type that performs 3D modeling while polymerizing macromolecules like inkjet. Modeling is performed by irradiating a tank filled with photo-cross-linkable resins with ultraviolet laser. Acrylic resin is similar to inkjet printing inks. Viscosity and surface tension in uncured state are related to molding accuracy. As a different feature, the material is contained in a liquid bath called a bathtub, and the structure is supported by the uncured liquid, so that a support structure is not required.

3. Examples of utilizing 3D printers

3D printers are widely used in industrial applications, while at the same time efforts are being made to utilize 3D printers in new fields. We will introduce numerous examples that have been done so far.

- 3D printer restaurant

Food Ink opened the world's first pop-up 3D print food restaurant. The food printer used in this project is a portable 3D printer "by Flow" developed by a startup company in the Netherlands. It is a 3D printer that can support various materials (PLA, ABS, polyethylene terephthalate (PET), nylon, bronze, silicon, food, etc.) by replacing the extruder. When used as a food printer, it is shaped by a method of extruding a food paste from a syringe pump.

- 3D printed pizza

Bee Hex has developed a 3D printer that outputs pizza. Pizza dough, tomato sauce, and cheese are made by dissolving the powder in water attached to the material tank of the printer. The material is extruded at high pressure and stacked from the pizza dough in turn. The basic structure is almost the same as the 3D printer, but the pressure mechanism and the print head are different. In addition, they are applying for a patent with technology to prevent dripping. Powder material was developed to accommodate the long-term stay of the space.

- 3D printed meal for elderly people

Biozoon developed a 3D printer powder mix for elderly and those with dysphagia. It is a material that solidifies when printed and dissolves when placed in the mouth as usable with general extruder-type 3D printers. This powder mix can form various shapes and it is possible to look and taste like real food using colorants.

- Food 3D printer that can output food in image form

XYZ Printing Japan has developed a food 3D printer "XYZ Food Printer." Traditional food 3D printers could output decorations to cookies but XYZ Food Printer is a new printer "output food in image shape."

- Introducing food 3D printer in fine restaurants

The Spanish fine restaurant "La Enoteca" has introduced a food 3D printer. Printed mashed potatoes in coral shape were developed. It is almost impossible to shape the mashed potatoes in the shape of corals with human hands. La Enoteca's expectations are that it can quickly respond when complex shaped orders come in.

- Gummy 3D printer

German confectionary manufacturer Katjes developed Gummy 3D printer "The Magic Candy Factory." We can make your own customized gummy from selfie and images in a few minutes. We can make original gummy by choosing your favorite one from 8 assorted flavors and 10 different shapes. The gummy liquid is squeezed out from the syringe warmed to 70–90°, and the gummy is stacked.

- Screw type chocolate

Confectionery store Grandujour has developed screw type chocolate. Chocolate is molded using molds made with 3D printers, and it looks like a real screw.

- Sugar 3D printer

MIT graduate students developed 3D printers of sugar at the robot workshop. The 3D printer can melt the sugar at 150°C and shape candy.

- Cake made with 3D modeling and 3D printing

Ukrainian artist Dinara Kasko made beautiful cake using 3D modeling and 3D printer. She shapes geometric configuration with FDM 3D printers, and made various cakes with silicon type model based on them.

- 3D printer capable of printing 100 types of meals

The Netherlands Organization for Applied Scientific Research (TNO) has developed a 3D printer that can provide 100 types of meals customized for 20 residents of German care facilities. Developed food 3D printers can make personalized foods according to individual needs.

4. YONEZAWA *Itadakimasu* Research Group

"YONEZAWA *Itadakimasu* Research Group" is conducting research and development of new food using YONEZAWA regional foods and 3D printer since 2014. *Itadakimasu* is the word Japanese people use before eating, to thank the ingredients, the people involved, and the nature. This research group is aimed at development and promotion of local economy through development of new food. Participating members are Yamagata University and some other local companies and universities.

We have suggested "The three points of Yonezawa *Itadakimasu* Research Group" and provided an idea based on those.

The three points of "YONEZAWA *Itadakimasu* Research Group" are:

1. Shape of food related to famous items of Yonezawa.
2. Yonezawa related ingredients.
3. Utilizing university facility such as 3D printer, laser cutter, or food gel to introduce the activities of the YONEZAWA *Itadakimasu* Research Group.

4.1. Food 3D printer "E-Chef"

"E-Chef" is a food 3D printer that can print a picture on a tablet or a cookie (**Figure 2**) [5]. Ink uses jam made from local fruits. E-Chef was exhibited at the local festival and is now in actual use. E-Chef was popular with children and well taken up in local newspapers. However, we found it difficult to introduce it in the restaurants as there are such problems as slow forming speed, costly, and difficulty in mastering. So, we should think of what type of food and which method we should make how to make with the 3D printer. This is described in next section.

4.2. Development of food using 3D printer

Yonezawa has famous food "ABC." A is Tateyama apple, B is Yonezawa beef, and C is Yonezawa carp. We developed new foods using these local foods and 3D printer. Developed foods are "3D Tateyama apple JELLY," "3D Tateyama apple RAKUGAN," "3D Yonezawa beef hot jelly," and "3D carp jelly." These are made with 3D printing mold using a FDM 3D printer. We introduce how to make 3D printing mold using 3D printer and features of each food.

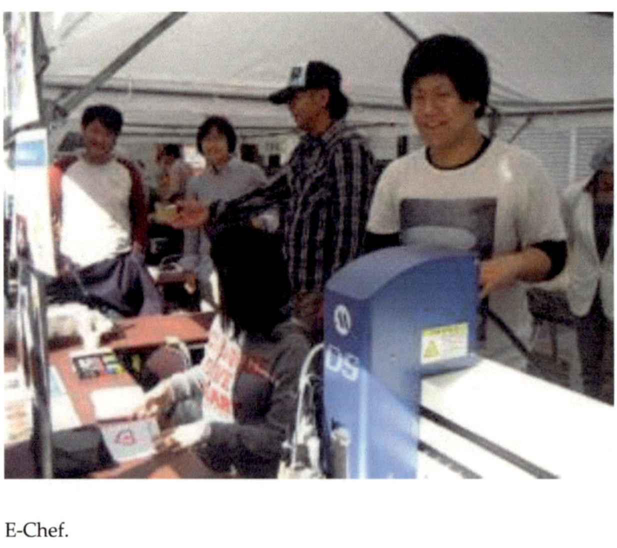

Figure 2. 3D printing by E-Chef.

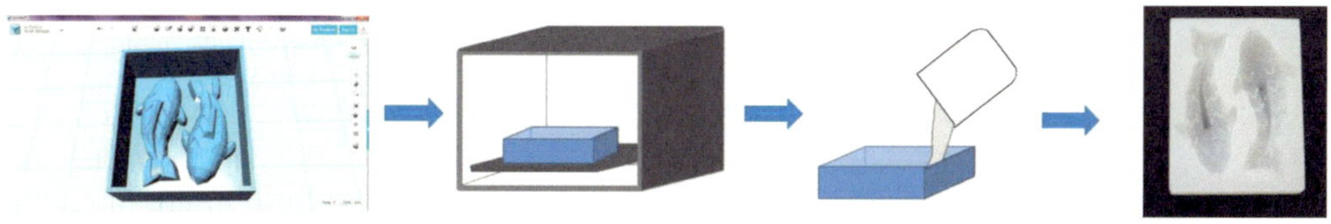

Figure 3. Manufacturing process of 3D printing mold.

4.2.1. Method of manufacturing 3D printing mold using 3D printer

3D printing mold of "3D Tateyama apple JELLY," "3D Yonezawa beef hot jelly," and "3D carp jelly" were made using FDM 3D printer. Stacking pitch was 0.2 mm and polylactic acid (PLA) was used as the filament. For making 3D printing mold, at first, we printed a cast. Then, food-grade silicone was poured into the cast to make a mold (**Figure 3**).

3D printing mold of "3D Tateyama apple RAKUGAN" was made with a FDM 3D printer and a laser cutter. RAKUGAN is a Japanese traditional confection made by adding a little moisture to the sugar and pushing it to the mold and drying it. Although the original mold is wooden, an acrylic plate processed with a laser cutter at the handle part, and parts for pushing the material were made with a 3D printer (**Figure 4**).

Figure 4. 3D printing mold of "3D Tateyama apple RAKUGAN".

Figure 5. 3D Tateyama apple JELLY.

4.2.2. 3D Tateyama apple JELLY

3D Tateyama apple JELLY is a jelly made from juice and powder of Taneyama apple (**Figure 5**). 3D printing mold is made from a 3D scan of a real Tateyama apple (**Figure 6**). The size of the jelly is 4.5 cm in diameter and 5.0 cm in height.

3D Tateyama apple JELLY is served as a dessert in the local Italian restaurant.

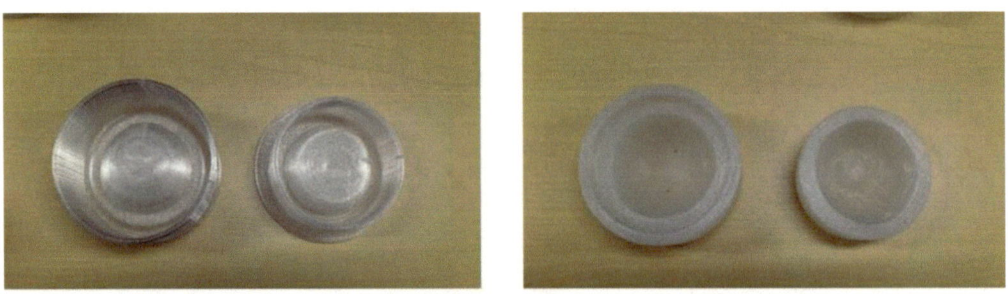

Figure 6. 3D Tateyama apple JELLY mold.

Figure 7. 3D Yonezawa beef hot jelly.

4.2.3. 3D Yonezawa beef hot jelly

"3D Yonezawa beef hot jelly" is made from Yonezawa beef consommé soup and agar (**Figure 7**). It is a rare jelly which is hard to dissolve when warmed, because it uses agar as a coagulant. Agar [1] made from seaweed usually forms into jelly at 0.7 wt.%. Jelly made from agar has a high melting temperature (68–84°C), keeps the shape in hot soup. 3D printing mold of the jelly has been designed in the motif of a snow lantern (**Figure 8**). The size of the jelly is 4.5 cm in length, 2.4 cm in width, and 1.7 cm in height.

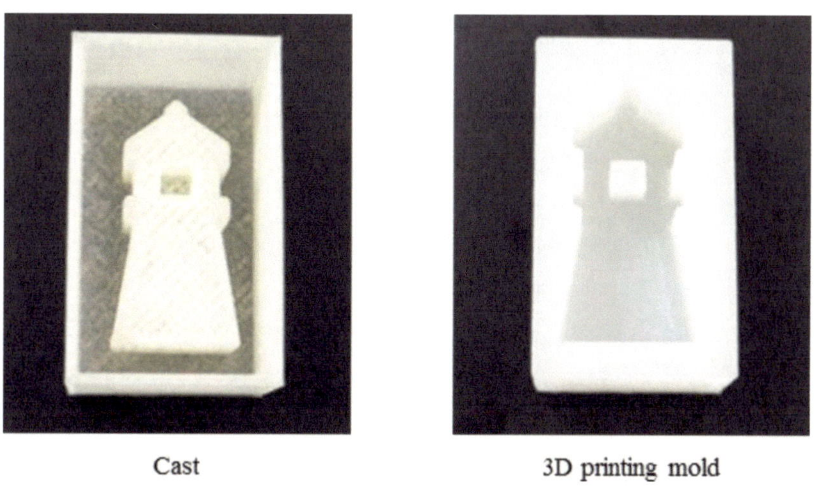

Figure 8. 3D Yonezawa beef hot jelly mold.

Figure 9. Restaurant menu 3D Yonezawa beef hot jelly.

"3D Yonezawa beef hot jelly" is served in the local Italian restaurant as appetizers or soup garnishes (**Figure 9**).

4.2.4. 3D carp jelly

3D carp jelly is a jelly in the shape of Yonezawa carp (**Figure 10**). "3D carp jelly" is served in the local carp restaurant as "3D carp hot pot." Jelly is made from soy milk, dried carp flakes, and gelatin. This Jelly melts easily and becomes soy milk soup when heated (**Figure 11**). Because the jelly contains heart-shaped konjac, konjac comes out when the jelly melts. Gelatin [2] made from collagen, contained in the skin of cows and pigs, usually forms into jelly at 2.0 wt.%. Jelly made from gelatin melts at room temperature (20–30°C). The size of the jelly is 6 cm in length, 2.7 cm in width, and 1.5 cm in height.

4.2.5. 3D Tateyama apple RAKUGAN

RAKUGAN is traditional Japanese confection made by adding a little moisture to the sugar and pushing it to the mold and drying it. RAKUGAN mold is originally wooden (**Figure 12**) and made

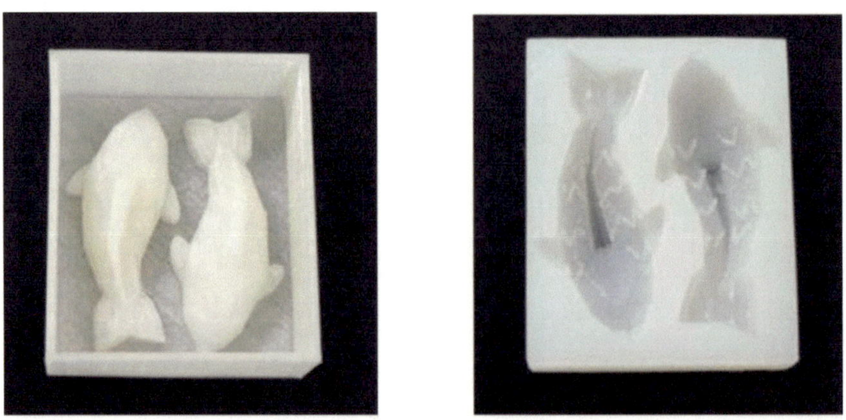

Cast 3D printing mold

Figure 10. 3D carp jelly mold.

Figure 11. Restaurant menu 3D carp hot.

by craftsmen, but the number of craftsmen are decreasing. Therefore, an acrylic board processed with a laser cutter, and parts for pushing the material were made with a 3D printer (**Figure 13**). 3D printing mold of RAKUGAN has been designed in the motif of a snow lantern. 3D Tateyama apple RAKUGAN is mainly composed of the powder of Tateyama apple and Wasanbon, which is fine-grained Japanese sugar (**Figure 14**). The size of the RAKUGAN is 3 cm in length, 1.5 cm in width, and 0.7 cm in height. 3D Tateyama apple RAKUGAN is sold at a local sweets shop.

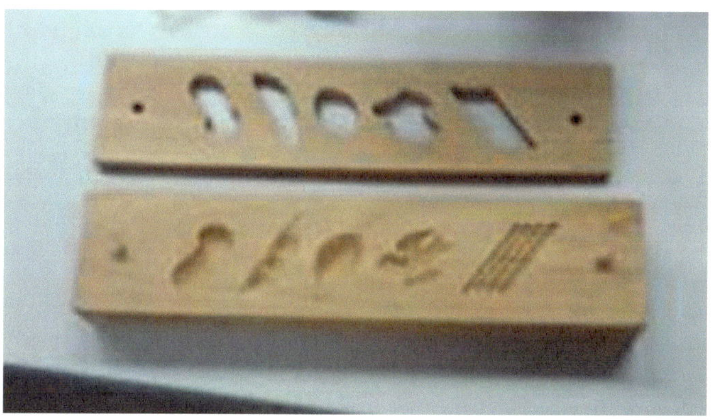

Figure 12. Wooden mold.

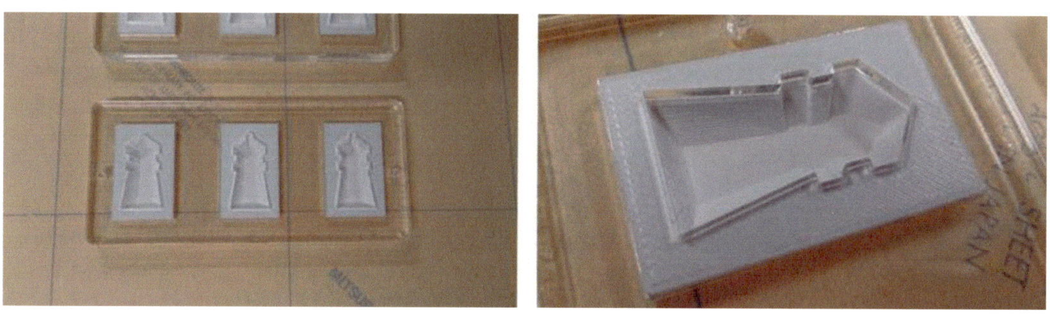

Figure 13. 3D printing mold of 3D Tateyama apple RAKUGAN.

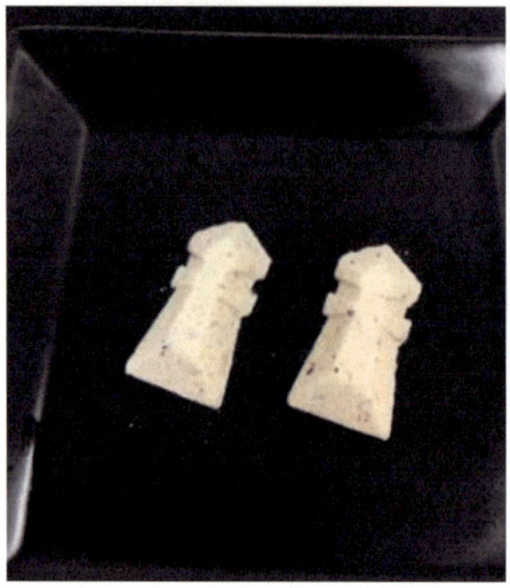

Figure 14. 3D Tateyama apple RAKUGAN.

5. Eki Fab

Eki Fab is a manufacturing space installed in the station building. 3D printers and laser cutters are installed, so that anyone can use them. Although this activity is open only every Saturday, many people use it when waiting for train at the station, including children and elderly people.

Eki Fab has held various events in the past. Among them, we have introduced events related to food.

5.1. Original jelly making class

We conducted an event "Let's make original jelly with 3D printer" for junior high school students. It is a part of the program of Japan Society for the Promotion of Science. The event was divided into three sections, (i) creation of jelly mold data, (ii) modeling a jelly mold with a 3D printer, and (iii) eating jellies made with original jelly mold.

i. Creation of jelly mold data

Jelly mold 3D data was created using AUTODESK 123D Design free CAD software. There are various operation methods for 3D CAD, and it is hard to master it. However, if it is a simple form like a jelly mold, it can be made with few steps so that data can be created in a relatively brief time. Junior high school students who participated in the event could finish creating jelly type data in about 2 h (**Figure 15**).

ii. Modeling a jelly mold with a 3D printer

Eki Fab has an FDM 3D printer. In FDM 3D printers, either filament of ABS or PLA is basically used. But we used a filament called PET+ which was approved by FDA (Food and Drug Administration) to touch foods (**Figure 16**).

iii. Eating jellies made with original jelly mold

Figure 15. Making 3D data with CAD.

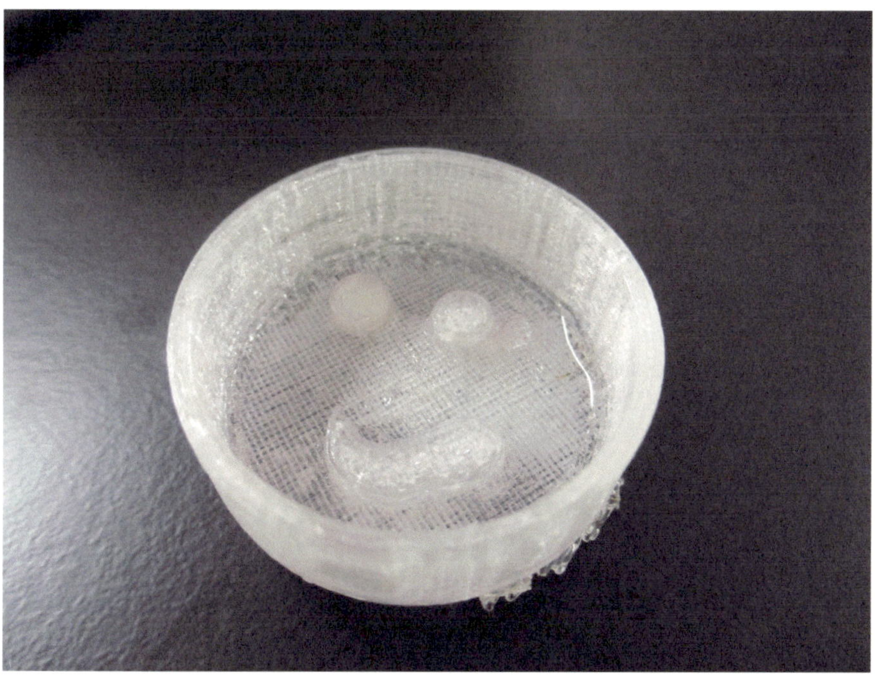

Figure 16. Jelly mold.

Figure 17. Eating jelly.

Even though most of the participants were seeing a 3D printer first time, everyone could create the original mold and make the jelly. Participants had a serious look during data creation and modeling, but a smile was seen when eating the finished original jelly (**Figure 17**).

6. Measurement of physical property of edible material for 3D food printer

We developed 3D food printer named "Food-y" after the E-Chef (**Figure 18**). Food-y is a printer developed jointly by MIRICE Co., Ltd., SEIKI Co., Ltd., and Yamagata University. In Food-y, shape of the food can be given by filling a syringe with a paste-like material such as bean paste or rice flour cookie and pressing it with a piston to eject the materials (**Figure 19**). However, other materials may not be able to shape depending on physical properties. Therefore, we have analyzed the texture and viscoelasticity of the material to ascertain conditions of lamina table material.

6.1. Materials for texture and viscoelasticity analyses

Materials were Mochi prepared by mixing rice flour and water and heating. Four Mochi were prepared by mixing 100, 125, 150, and 200 ml of water with 50 g of rice flour (Mochi 33 (rice flour 33%), Mochi 28 (rice flour 28%), Mochi 25 (rice flour 25%), and Mochi 20 (rice flour 20%), respectively).

http://dx.doi.org/10.5772/intechopen.70652

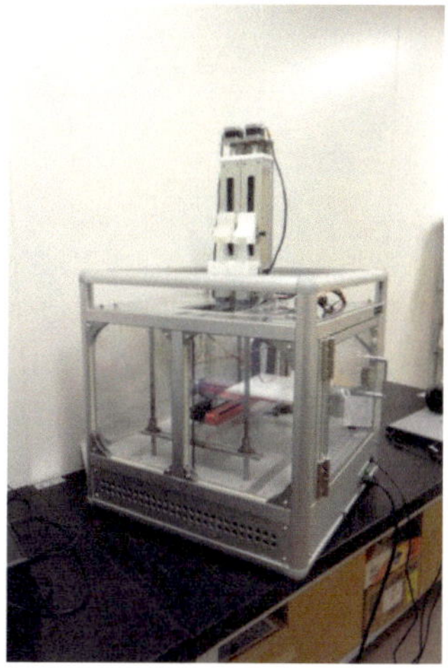

Figure 18. 3D food printer "Food-y".

6.2. Modeled object and volume

Modeling was carried out at the material temperature kept at 25°C. **Figure 20** shows the 3D data of the apple recessed in the center used for molding. **Figure 21** shows what was molded in Food-y. **Table 1** shows the volume of the modeled objects.

6.3. Measurement of texture

Food has characteristics of texture. The texture as a dynamic property includes factors such as hardness, cohesiveness, and adhesiveness. Cohesiveness represents the easiness of cohesion

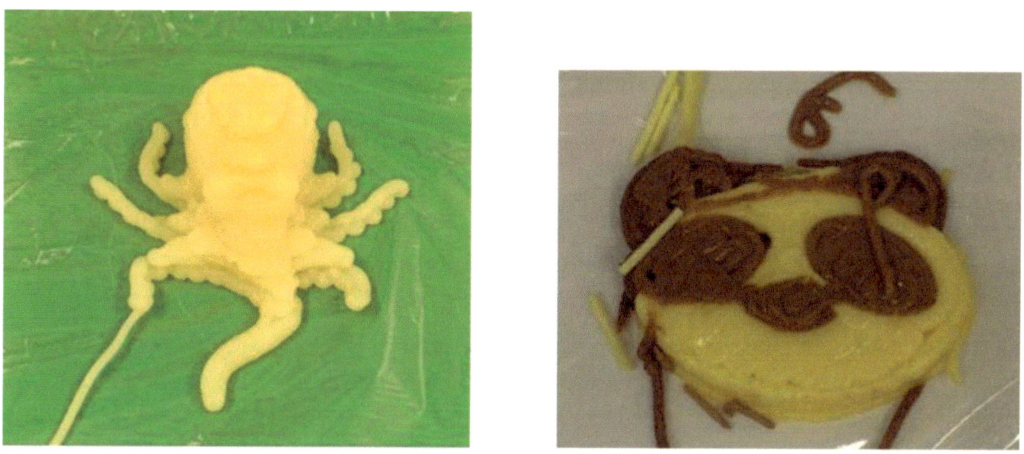

Bean paste octopus　　Rice flour cookie panda

Figure 19. Bean paste and rice flour cookie by 3D printed.

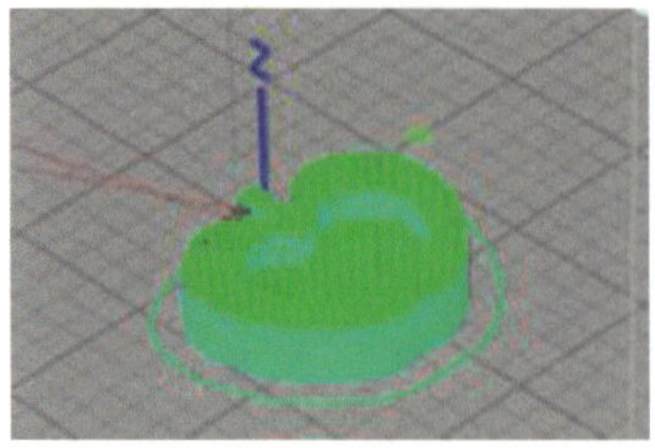

Figure 20. 3D data of apple.

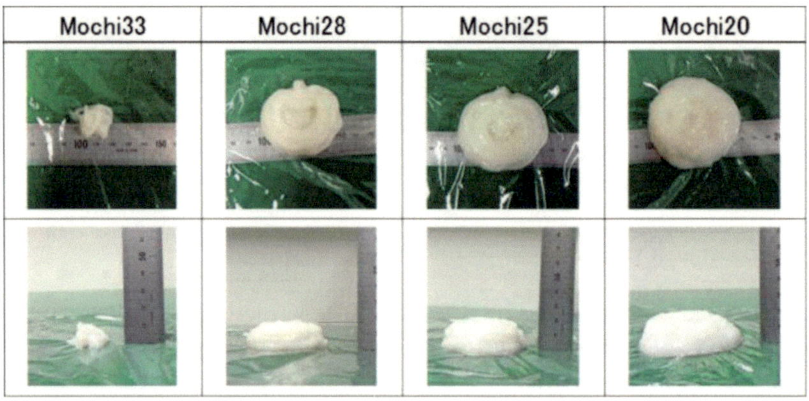

Figure 21. Printed Mochi by Food-y.

when the food is eaten, and adhesiveness represents the force to attach and separate from teeth, tongue, and oral cavity when the food is eaten [3].

We measured the texture before modeling for Mochi 33, 28, 25, 20. We used EZ Test (SHIMADZU CORPORATION) (**Figure 22**) as a texture analyzer. For the test, hardness, cohesiveness, and adhesiveness were measured nine times for each sample, and the results are shown in **Figures 23–25**, respectively.

6.4. Measurement of viscoelasticity

Mochi has a viscoelastic body. Therefore, we evaluated the workability of Mochi by measuring viscoelasticity. Measurement of viscoelasticity by a rheometer is technically established, and it can be said that it is appropriate because it is excellent in quantitative aspects [4].

For the Mochi 33, dynamic viscoelasticity measurement in the linear region was carried out as measurement of viscoelasticity before modeling. We used MCR 301 (Anton Paar Co., Ltd.) (**Figure 26**) as a rheometer. Cone plate ϕ 25 mm Angle 2 degrees and ϕ 50 mm Angle 1 degree

Sample	**Mochi 33**	**Mochi 28**	**Mochi 25**	**Mochi 20**
Volume (ml)	2	21	43	44

Table 1. Volume of the modeled objects.

http://dx.doi.org/10.5772/intechopen.70652

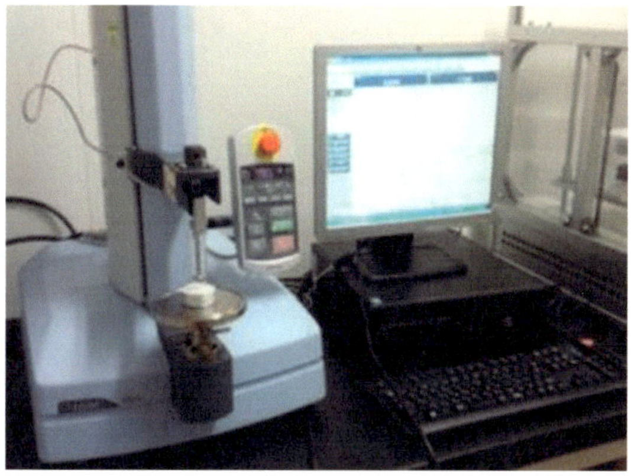

Figure 22. Texture analyzer EZ Test.

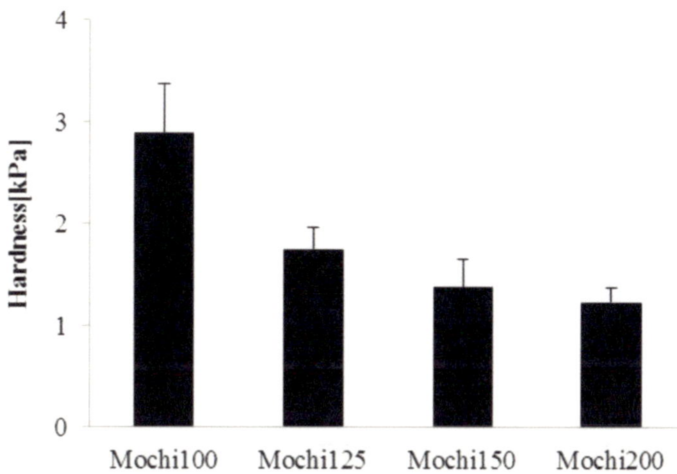

Figure 23. Hardness of Mochi.

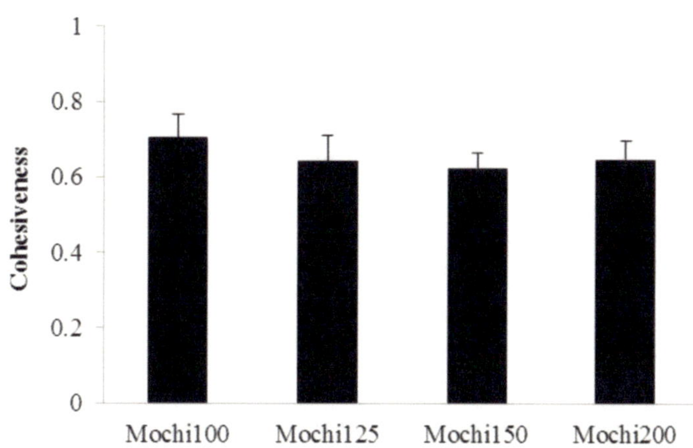

Figure 24. Cohesiveness of Mochi.

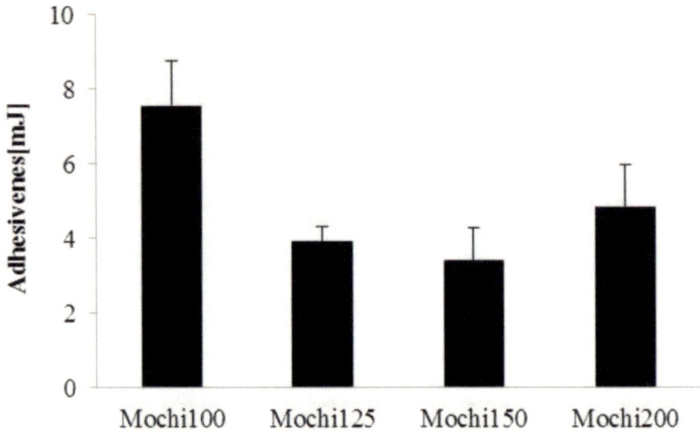

Figure 25. Adhesiveness of Mochi.

were used (**Figure 27**). The results of storage elastic modulus G′ and loss elastic modulus G″ are shown in **Figure 28**. Also, storage elastic modulus G′ and loss elastic modulus G″ at an angular frequency of 1 rad/s and the hardness measured by a texture analyzer are shown in the same graph (**Figure 29**).

6.5. Results

Mochi 28 could make beautiful shapes enough to easily understand the shape of the apple (**Figure 21**). Therefore, consider Mochi 28 as a reference. Mochi 33 is deficient in ejection and Mochi 25 and Mochi 20 are ejected excessively (**Table 1**).

Hardness decreases as the proportion of rice flour becomes smaller (**Figure 23**). Cohesiveness is almost constant regardless of the sample (**Figure 24**). Adhesiveness is variable (**Figure 25**).

G′ and G″ becomes smaller as the amount of rice flour decreases (**Figure 28**).

From Mochi 33 to Mochi 25, hardness, G′ and G″, tends to be similar (**Figure 29**).

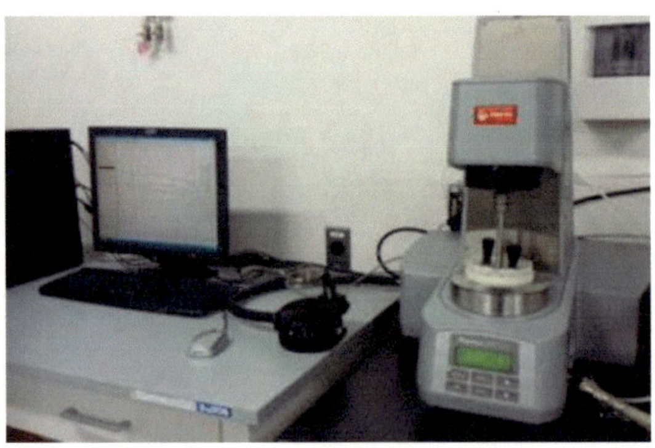

Figure 26. Rheometer MCR301.

http://dx.doi.org/10.5772/intechopen.70652

Figure 27. Cone plate of rheometer.

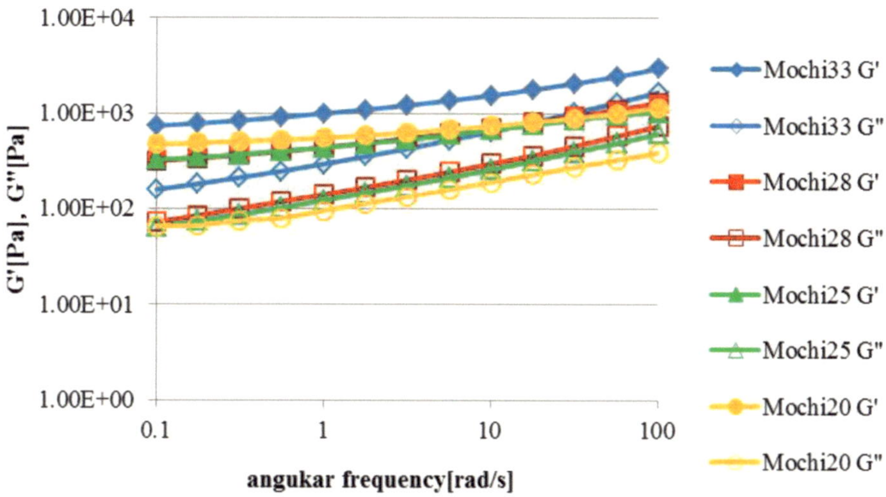

Figure 28. G′ and G″ of Mochi.

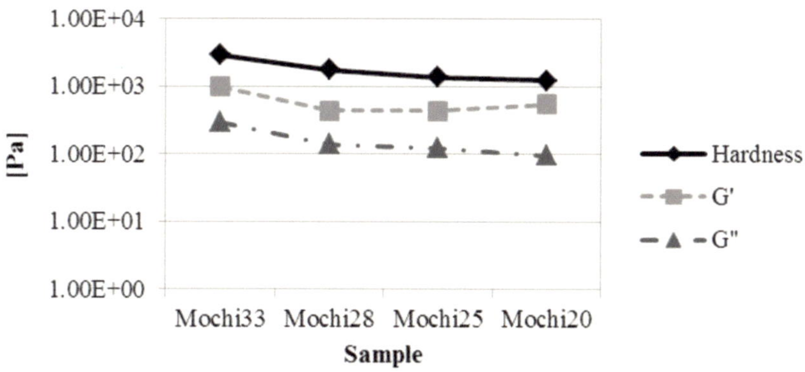

Figure 29. Hardness, G′, and G″ of Mochi.

6.6. Consideration

From the texture test, we can see that the sample is hard if volume is small and if the sample is soft volume becomes excessive. Easiness of 3D laminate modeling can be determined from the hardness measured by the texture analyzer.

From the comparison of the texture test and the viscoelasticity test, the hardness of the storage elastic modulus and the loss elastic modulus of the sample tend to be similar.

6.7. Conclusion

The simplicity of 3D laminate modeling by food-3D printers can be determined by the hardness measured by a texture analyzer. The texture analyzer is standardized in the laboratory of food and considered to be useful as a discrimination method. In addition, the hardness measured by the texture analyzer agrees well with the viscoelasticity measured by the rheometer. In the laboratory of soft material engineering, it is possible to evaluate the viscoelasticity by a rheometer and advance the research to predict the ease of 3D laminate modeling.

Future work will involve conducting similar research on food samples other than Mochi and investigate the generality of the above conclusion. We will also develop a method to judge easiness of 3D laminate modeling from measured values of viscoelasticity. This will ultimately establish a methodology for finding the optimum 3D lamination condition from the physical properties of soft materials.

7. 3D printing and future foods

As discussed, there are many applications of 3D printers in the field of food. Common to all cases is that the producer is using 3D printers considering "appearance" of food. "Appearance" of food is very important; it is one of the key factors to make delicious meals. Especially, it is said that care foods are so soft that they look bad and it also affects the amount of meals. Therefore, using a 3D food printer or 3D printing mold, "looks" can be greatly improved.

3D food printing has many challenges, such as shortage of edible material, less-accuracy during modeling, and difficulty in real use. However, we can expand the possibilities of food shaping by advancing what we can do now such as make a 3D printing mold with 3D printer etc.

One of the features of 3D printers is that materials can be shaped if there is only 3D data. As the development of food 3D printers progresses, it may be possible to adjust the shape, taste, nutrition, etc. according to individual's physical condition automatically based on the digital data of the food, and cooking can be done with one click of operation.

Author details

Mai Kodama[1]*, Yumiko Takita[1], Hideaki Tamate[1], Azusa Saito[1], Jin Gong[2], Masato Makino[1,3], Ajit Khosla[1,3], Masaru Kawakami[1,3] and Hidemitsu Furukawa[1,2,3]

*Address all correspondence to: maikodama@yz.yamagata-u.ac.jp

1 Soft & Wet matter Engineering Lab (SWEL), Life-3D Printing Innovation Center (LPIC), Yamagata University, Yamagata, Japan

2 Department of Systems Innovation, Faculty of Engineering, Yamagata University, Yamagata, Japan

3 Department of Mechanical Systems Engineering, Graduate School of Science and Engineering, Yamagata University, Yamagata, Japan

References

[1] Uzuhashi Y, Taki C. Kinds, properties and usage of agar. Journal of Cookery Science of Japan. 2005;**38**(3):292-297. DOI: http://doi.org/10.11402/cookeryscience1995.38.3_292

[2] Nagatsuka N, Ohno T, Okawa Y, Kawamura F, Nagao K. Factors affecting the gelatin of a compound gelatin solutions—Effects of adding sugar and alcohol. The Japan Society of Cookery Science. 2003;**36**(4):364-370. DOI: http://doi.org/10.11402/cookeryscience1995.36.4_364

[3] Nishinari K. Physical properties of food and problems in measurement. The Japan Society of Home Economics. 2013;**64**(12):811-822. DOI: http://doi.org/10.11428/jhej.64.811

[4] Miura M. Fascination of food rheology. 4th lecture: Viscoelasticity of foods. Journal of the Society of Rheology, Japan. 2015;**43**(2):47-50. DOI: http://doi.org/10.1678/rheology.43.47

[5] Shitara M, Gong J, Serizawa R, Makino M, Nishioka A, Furukawa H. Food creation with 3D gel printer for foods. Transactions of the JSME (in Japanese). 2015;**2015**:15-00008. DOI: http://doi.org/10.1299/transjsme.15-00008